Tyler's Herbs of Choice

The Therapeutic Use of Phytomedicinals

Third Edition

Dennis V.C. Awang

CRC Press
Taylor & Francis Group
Boca Raton   London   New York

CRC Press is an imprint of the
Taylor & Francis Group, an **informa** business

CRC Press
Taylor & Francis Group
6000 Broken Sound Parkway NW, Suite 300
Boca Raton, FL 33487-2742

© 2009 by Taylor & Francis Group, LLC
CRC Press is an imprint of Taylor & Francis Group, an Informa business

No claim to original U.S. Government works
Printed in the United States of America on acid-free paper
10 9 8 7 6 5 4 3 2

International Standard Book Number-13: 978-0-7890-2809-9 (Hardcover)

Library of Congress Cataloging-in-Publication Data

Awang, Dennis V. C.
    Tyler's herbs of choice : the therapeutic use of phytomedicinals / Dennis V.C. Awang. -- 3rd ed.
        p. cm.
    Rev. ed. of: Tyler's herbs of choice / James E. Robbers, Varro E. Tyler. c1999.
    Includes bibliographical references and index.
    ISBN 978-0-7890-2809-9 (hardcover : alk. paper) -- ISBN 978-0-7890-2810-5 (pbk. : alk. paper)
    1. Herbs--Therapeutic use. I. Robbers, James E. Tyler's herbs of choice. II. Title.

RM666.H33R6 2009
615'.321--dc22   2009006363

Visit the Taylor & Francis Web site at
http://www.taylorandfrancis.com

and the CRC Press Web site at
http://www.crcpress.com

Dedication

This second revision of Professor Tyler's original text is dedicated to his memory. The late Dr. Varro E. "Tip" Tyler (1926–2001) was a dear friend, a consummate gentleman, and, undoubtedly, one of the most outstanding botanical medicine scientists of the twentieth century. His contributions to academia and to public education, as well as his tireless efforts in promoting sensible, effective regulation of the herbal industry, are an everlasting monument to his scientific preeminence and his generous character.

Contents

Foreword

The growth in consumer use of herbal preparations for health-related purposes has generated continuing pressure on educational institutions to train conventional health practitioners in this burgeoning area. Now, more than ever before, pharmacists, physicians, nurses, dietitians, and other professionals are being confronted with situations in which their clients are using a wider variety of dietary supplements for an increasing range of health conditions. Unfortunately, these health professionals often lack adequate training and skills required to assess the overall safety and effectiveness of dietary supplements properly. Although varied and numerous print and electronic resources are available to help guide health care practitioners in understanding the applications of many herbal and other dietary supplements, there is still a need for basic instruction in this area.

In my own experience at the American Botanical Council (ABC, a non-profit herbal education organization), the many pharmacy student interns who come for six-week clinical rotations as part of their fulfillment of the requirements to complete training to become a doctor of pharmacy have no formal training in herbs, phytomedicines, or related subjects. This is certainly also true of the dietitian interns who participate in much shorter, two-week rotations.

Would that they had read and studied this book as part of their professional education! Their patients would benefit greatly from their enhanced basic understanding of the role that herbs and phytomedicinal products can play in both self-care and health care. Health professionals, particularly pharmacy and medical students, will find that this book is one of the most authoritative introductions to herbs and phytomedicines available in the English language. It is particularly useful as an introductory training manual.

What makes *Tyler's Herbs of Choice* so authoritative and reliable is its initial author, the late Professor Varro E. Tyler, a true giant in the field of pharmacognosy and pharmacy education. Its authority is also supported by the two authors who revised the subsequent editions.

Pharmacognosy is the study of drugs of natural origin, whether derived from plants or animals. Until about the 1970s and 1980s, when it was gradually replaced by medicinal chemistry, pharmacognosy was a required course for all pharmacy students. In pharmacognosy, one would study the plant origins of drugs, where the plants were originally harvested or cultivated, which plant parts contained the primary active constituents, how these constituents were extracted from the plant materials, their pharmacology and toxicology, the microscopy required to identify powdered vegetable drugs, and the chemical tests available to determine the identity of various drug plants. This knowledge was a necessary part of a pharmacist's education from before the turn of the twentieth century until at least the 1960s or 1970s.

During the 1940s through the 1970s, herbal products—many of which were formerly recognized as official medicines in the *United States Pharmacopeia*—were systematically removed from modern medicine and pharmacy. This decline occurred not because plant preparations were found to be unsafe or ineffective but, rather, primarily because they fell into disuse in conventional medicine and pharmacy, which began to rely increasingly on synthetic, single-chemical pharmaceutical medicines. However, as the use of chemically complex botanical medicines declined in conventional medicine, some of these same plant materials and many others began to migrate to health food stores in numerous forms—as herbal teas, tinctures, capsules, tablets, and other forms.

During the past few decades, millions of Americans, as well as people all over the world, have expressed considerable interest in, and even preference for, natural medicinal products, which go by a number of different names (e.g., herbal remedies, herbal medicines, botanical medicines, herbal medicinal products, traditional medicines, folkloric medicines). Today, this preference for herbal remedies has expanded to an economically significant sector; in 2005, the sales of botanical remedies in the United States alone was more than $4.4 billion.[1] In some countries, particularly Germany and other European nations, the term *phytomedicine* is preferred; the term usually refers to a plant extract—often "standardized" to contain a guaranteed range of active or marker chemical compounds—with at least a modicum of documented safety and efficacy.

For more than twenty years Professor Tyler acted as dean of the School of Pharmacy at Purdue University, where he taught a course in pharmacognosy. (He later became vice president of academic affairs and provost at Purdue before his retirement.) Professor Tyler was known and respected nationally and internationally as a premier figure in pharmaceutical education, particularly in pharmacognosy. He was the senior author of the leading textbook in this field,[2] four editions of which were used by an entire generation of American pharmacists from the 1960s through the

1980s (i.e., to the extent that a college of pharmacy still offered pharmacognosy in the 1980s).

Eschewing the use of e-mail, Tyler provided faxes to his colleagues with his scrupulously considered editorial corrections, suggestions, comments, etc. He was meticulous with a manuscript, catching every misspelled word, incorrect phrase, comma splice, and other grammatical problems that often go undetected in contemporary publications.

For six years I taught a course called "Herbs and Phytomedicines in Today's Pharmacy" in the College of Pharmacy at the University of Texas at Austin. I employed the second edition of this book, the one revised by Professor James E. Robbers, a student and colleague of Professor Tyler's, as the primary textbook for my students. Their readings were supplemented with some of the American Botanical Council's books, as well as numerous other readings available in ABC's HerbClip™, now comprising more than thirty-three hundred summaries and critical reviews of the clinical and scientific literature on herbs.

While reviewing various chapters of *Herbs of Choice* for the forthcoming week's three-hour lecture, I would sometimes come across statements in the book that I considered inadequate or incomplete in light of more recent advances in the scientific and clinical literature or recent regulatory developments. In some cases, the range of herbs and herbal preparations discussed by Professor Tyler did not include some herbal preparations that had been subsequently introduced into the North American market and for which a growing body of scientific and clinical data supported safety and probable efficacy.

At any rate, the dietary supplement market, as well as the clinical literature, was growing and changing. To enable pharmacy students to maintain their grasp on the types of products about which retail pharmacy consumers were anticipated to inquire, *Tyler's Herbs of Choice*, like any book, needed some revision.

Varro Tyler had a deep fascination and respect for German phytomedicine, the quality control used by the manufacturers, the rational regulation by the government (Commission E), the inclusion of phytomedicines in pharmacy and medical school education, and their rational use in pharmacy and medicine as both prescription and nonprescription drugs. He graciously assented to write the forewords to the American Botanical Council's first two books: *The Complete German Commission E Monographs: Therapeutic Guide to Herbal Medicines* (1998) and *Herbal Medicine—Expanded Commission E Monographs* (2000) because he was the driving force that motivated us to produce a complete translation and cross-referencing of the now widely cited monographs from the German special expert committee on herbs and phytomedicines (commonly referred to as the Commission E).

Professor Tyler and I were scheduled to lead a tour for American pharmacists and physicians of herbs and wines of the Rhine in August 2001. Just back home after a visit in Austria with his wife and partner Virginia, he died suddenly on August 22, 2001—a sad day for the worlds of pharmacognosy and herbal medicine.

On several occasions, I contacted Bill Cohen of Haworth Press to suggest that Tyler's book be revised. Although I sometimes fantasized that I might be the person for the task because I was so familiar with its contents as well as its primary author, I was genuinely heartened and gratified to learn that my good friend and colleague, Dr. Dennis V. C. Awang, had been asked to do so. With respect to the herb and medicinal plant literature, there is no more knowledgeable expert and no more punctilious editor of this subject in all of North America. I have known him for more than twenty-five years—ever since he was the key medicinal plant expert in the Canadian government's former Health Protection Branch, equivalent to the U.S. Food and Drug Administration.

I know of only one other expert in the field of pharmacognosy and natural products who employs the level of scrutiny that Professor Tyler used to bring to his editorial activities: the author and editor of this revised book, Dr. Awang. As a friend and colleague of Tyler's, Dr. Awang shared his passion for science, natural products chemistry, and phytomedicines. Both would relish the opportunity to get out their red pens and correct errors in a manuscript intended for publication in a botany, pharmacy, medical, or phytomedicine journal. This newly revised book has Dr. Awang's fingerprints all over it, and he is a rightful successor to Tyler. Both employ the English language in a highly accurate manner. Awang, one of North America's most respected and knowledgeable natural products chemists, is also a scrupulous editor.

About fifteen years ago, when Dr. Awang was peer-reviewing a manuscript for publication in ABC's journal *HerbalGram,* he sent some remarks that were highly critical of the author's apparent lack of knowledge of proper botanical taxonomy. When I confronted him on whether he, as a chemist, was adequately qualified to make such a strong taxonomic criticism, he replied with what I have often referred to as a quintessential Awangian truism: "My dear Mark," he responded in his impeccable Queen's English, "everyone knows that it's much easier for a chemist to learn botany, than for a botanist to learn chemistry!"

This book is neither about botany per se nor about natural products chemistry, although it includes some elements of both disciplines. This book combines the various scientific fields of herbal medicine, phytomedicine, and pharmacognosy with the modern clinical trials that in general continue to support the compelling rationale for using properly prepared plant products in self-care and health care. Whether you are a pharmacy or medical student; a pharmacist, physician, or other health professional;

or perhaps an interested layperson seeking an authoritative introduction to the vast and highly interesting field of botanical medicine, there is no doubt that you will find significant value in these pages.

Mark Blumenthal
Founder and executive director, American Botanical Council, Austin, Texas; editor, HerbalGram and HerbClip; senior editor, The Complete German Commission E Monographs—Therapeutic Guide to Herbal Medicines and The ABC Clinical Guide to Herbs

References

1. Blumenthal, M., G. K. L. Ferrier, and C. Cavaliere. 2006. Total sales of herbal supplements in United States show steady growth. *HerbalGram* 71:64–66.
2. Tyler, V. E., L. R. Brady, and J. E. Robbers. 1988. *Pharmacognosy,* 9th ed. Philadelphia, PA: Lea & Febiger.

Preface

Since publication in 1999 of the first revision of *Herbs of Choice*, a plethora of publications has extended both the range and depth of herbal medicinal science. Notable has been the introduction to the West of a number of herbs long popular in Eastern traditional medicine systems. Prominent among these have been *Andrographis paniculata, Petasites hybridus* (butterbur), *Centella asiatica* (gotu kola), *Bacopa monnieri,* and *Citrus aurantium* (bitter orange) as an ephedra substitute.

There has also been an explosion of concern and scientific investigation into the potential for herb–drug interaction influencing clinical outcomes—from grapefruit (*Citrus x paradisi*) to St. John's wort (*Hypericum perforatum*). Examination of bleeding and coagulation associated with herbs has been expanded. Recent research has also clarified certain aspects of the mechanisms of action of a number of popular herbs, such as feverfew, ginkgo, and ginseng. Particularly, a wholesale revision of the feverfew treatment has been effected. Phytochemical treatment of liver disease and the activity of phytoestrogens have been more widely explored.

The expansion of herbal research in the United States, heralded in the second edition of *Herbs of Choice*, has markedly increased. Since the 1997 publication of a fair and fairly positive assessment of *Ginkgo biloba* extract for treatment of Alzheimer's disease and vascular dementia in the *Journal of the American Medical Association* (*JAMA*), a number of articles have appeared in medical journals, some blatantly deficient in appreciation of herbal scientific parameters. Conspicuous among the latter have been two *JAMA* reports[1,2] regarding the effectiveness of St. John's wort (SJW) in major depression, without acknowledging that SJW is recognized for its effectiveness in treating mild to moderate depression. The results of a later study were widely publicized as demonstrating the ineffectiveness of SJW, without noting that the widely prescribed pharmaceutical sertraline was also not significantly different in effect from placebo. Most herbal scientists attribute the failure of both test treatments to relatively long-term, chronic serious depression.

The effectiveness of echinacea in treatment of the common cold has also come under scrutiny by the medical community: Knight's critique[3] focused on two studies that generated negative findings but addressed *prevention* rather than *symptomatic* treatment of rhinoviral inoculation. A recent article in the *New England Journal of Medicine* (*NEJM*) found no evidence of any clinically significant efficacy of three extracts of *Echinacea angustifolia* root.[4] The results of that study prompted Samson to criticize the National Center for Complementary and Alternative Medicine for funding investigation of "implausible remedies" and charge a prevalent tendency of herbal enthusiasts to "dismissing disproof."[5] Two other recent *NEJM* publications report ineffectiveness of three popular herbal remedies: saw palmetto for benign prostatic hyperplasia,[6] glucosamine-chondroitin for painful knee osteroarthritis,[7] and black cohosh for hot flashes.[8]

The herbal scientific literature is plagued by mixed results from different studies with a number of medicinal plants. These mixed results are widely believed to be due to methodological inadequacies or variation in the chemical character of test treatments.

A recent publication lamented the lack of adequate characterization of herbal supplements subjected to randomized controlled trials (a position long embraced by the late Professor Tyler[9]) and stressed the importance of quality control issues in ensuring "the value of otherwise well-designed clinical trials."[10] Yet another publication assessed five popular botanical dietary supplements from nine manufacturers; echinacea, ginseng, kava, SJW, and saw palmetto were analyzed for compliance of marker compound content with label claims.[11] Although little variability was noted between different lots of the same brand, content of marker compound varied widely between brands, as did information regarding serving recommendations and herbal parameters such as species, part of plant, and marker compounds targeted.

All these considerations recommend much closer attention to all phases of the manufacture of botanical test preparations and committed response to the near-mantra call for larger, better-designed clinical trials of longer duration, with properly characterized treatments.

I am considerably indebted to Kenneth Jones, Armana Research, BC, for contributions to my scientific information base and for thoughtful and critical discussions of a wide range of herbal medicinal issues. The American Botanical Council's HerbClip™ service has also been a valuable source of current publications.

Dennis V. C. Awang
White Rock, BC, Canada

References

1. Shelton, R. C., M. B. Keller, A. Gelenberg, D. L. Dunner, R. Hirschfeld, M. E. Thase, et al. 2001. *Journal of the American Medical Association* 285 (15): 1978–1986.
2. Hypericum Depression Trial Study Group. 2002. *Journal of the American Medical Association* 287:1807–1814.
3. Knight, V. 2005. *Clinical Infectious Diseases* 40:807–810.
4. Turner, R. B., R. Bauer, K. Woelkart, T. C. Hulsey, and J. D. Gangemi. 2005. *New England Journal of Medicine* 353:341–348.
5. Sampson, W. 2005. *New England Journal of Medicine* 353:337–339.
6. Bent, S., C. Kane, K. Shinohara, J. Neuhaus, E. S. Hudes, H. Goldberg, and A. L. Avins. 2006. *New England Journal of Medicine* 354:557–556.
7. Clegg, D. O., D. J. Reda, C. L. Harris, M. A. Klein, J. R. O'Dell, M. M. Hooper, et al. 2006. *New England Journal of Medicine* 354:795–808.
8. Newton, K., S. D. Reed, L. Grothaus, K. Ehrlich, J. Guiltinan, E. Ludman, and A. Z. Lacroix. 2005. *Maturitas* 16:134–146.
9. Tyler, V. E. 2000. *Scientific Review of Alternative Medicine* 4 (2): 17–22.
10. Wolsko, P. M., D. K. Solondz, R. S. Phillips, S. C. Schachter, and D. M. Eisenberg. 2005. *American Journal of Medicine* 118:1087–1093.
11. Krochmal, R., M. Hardy, S. Bowerman. Q.-Y. Lu, H.-J. Wang, R. M. Elashoff, and D. Heber. 2004. *Evidence-Based Complementary and Alternative Medicine* 1 (3): 305–313.

The author

Dennis V. C. Awang, FCIC (fellow of the Chemical Institute of Canada, 1988), is a graduate of Queen's University, Kingston, Ontario (BSc, 1960; PhD, 1967). He conducted postdoctoral studies in organic chemistry at the Universities of Michigan and Illinois. Prior to establishing his own consulting business, Dr. Awang was employed as a research scientist at Health and Welfare Canada for 24 years. At the Bureau of Drug Research of the Health Protection Branch, he directed research in support of the regulatory bureaus of the Drugs Directorate in the areas of drug stability and methodology development for antibiotics, hormones, and natural products. He was also the official spokesperson in herbal science for the Canadian government for many years.

The author of more than 150 research and scientific publications, Dr. Awang was on the editorial advisory board of the *Journal of Herbs, Spices and Medicinal Plants* and the *Alternative Therapies in Women's Health* newsletter. He is a member of the international editorial board of FACT (Focus on Alternative and Complementary Therapies) and a contributing editor to *HerbalGram,* the organ of the American Botanical Council (ABC); for many years, he was one of the regular authors of the *Canadian Pharmaceutical Journal's Herbal Medicine* series of reviews, and a regular lecturer and media commentator on medicinal plant and natural products science. Dr. Awang is a member of the ABC's advisory board; he has also served on the USP Advisory Panel on Identification and Standardization of Natural Products and the faculty of Columbia University's annual course of herbal medicine for physicians and surgeons, the Botanical Advisory Committee of Leiner Health Products, and as a director of Chai-Na-Ta Corp., the world's largest American ginseng grower. He is also a member of the American Chemical Society.

Dr. Awang served for 2 years on the Advisory Committee of the University of Illinois at Chicago/U.S. National Institutes of Health (NIH) Botanical Dietary Supplements Research Center (BDSRC), and as a chair of the advisory board of the NIH BDSRC of Purdue University–University of Alabama (Birmingham). In 1994, he was granted the Martin de la Cruz award by the Mexican Academy of Traditional Medicine for distinguished contribution to medicinal plant research. Dr. Awang was recently appointed North American, co-editor of the international journal *Phytomedicine*.

chapter one

Basic principles

Definitions

Herbs are defined in several ways, depending on the context in which the word is used. In botanical nomenclature, the word refers to nonwoody seed-producing plants that die to the ground at the end of the growing season. In the culinary arts, it refers to vegetable products used to add flavor or aroma to food. But in the field of medicine, the term has a different, yet specific, meaning. Here it is most accurately defined as *crude drugs of vegetable origin utilized for the treatment of disease states, often of a chronic nature, or to attain or to maintain a condition of improved health.* Pharmaceutical preparations made by extracting herbs with various solvents to yield tinctures, fluidextracts, extracts, or the like are known as phytomedicinals (plant medicines).

In the United States, the choice of an herb or phytomedicine for therapeutic or preventive purposes is usually carried out by the patient. To state it differently, the herb is self-selected because physicians here are not ordinarily educated in the use of such medicinals. However, in many other countries, herbs and phytomedicinals are prescribed by doctors with considerable frequency.

It cannot be emphasized strongly enough that herbs in their medicinal sense are drugs. For reasons that will become apparent when the legal concerns regarding herbs are discussed in the Appendix, certain special-interest groups continually emphasize their point of view that medicinal herbs are foods or dietary supplements. Scientifically (although not legally under current law), that is not the case. If they are used in the treatment (cure or mitigation) of disease or improvement of health (diagnosis or prevention of disease), they conform to the definition of the word *drug.*

However, a limited number of botanical products may qualify as foods or drugs or both, depending on the specific intent of the user. Employed as a flavoring agent in cooking, garlic is clearly a food. When it is used to control hypertension or excessively high cholesterol levels, it is a drug. Possibly, some may use it for both purposes at the same time. The millions of elderly Americans who drink prune juice daily probably consider it both a pleasant breakfast beverage (food) and a means of maintaining "regularity" (drug). Psyllium (plantago) seed also falls in this ambivalent category. Widely used on a daily basis as an over-the-counter (OTC) bulk laxative that can be purchased without a physician's prescription, the seed would be classified by most users as a drug. The same nutritious seed incorporated into a breakfast cereal is certainly nothing more than a healthful food.

The only way to settle this dilemma is to admit that relatively few plant products defy precise definition as either foods or drugs. Fortunately, in the vast panoply of herbs, relatively few present this classification problem. Those that do are mostly specialized storage organs of the plants, such as fruits, seeds, or fleshy underground parts rich in carbohydrates. The problem is much less frequently encountered with the flowering tops, leaves and stems, barks, rhizomes, or roots that constitute most herbs. The basic definition still applies. Herbs used for medicinal purposes are drugs.

Differences between herbs and other drugs

Herbs are different in several respects from the types of purified therapeutic agents we have become accustomed to call drugs in the last half of the twentieth century. In the first place, they are more dilute than the concentrated chemicals that are familiar to us in the form of aspirin tablets or tetracycline capsules. A simple example will illustrate the difference. One can take caffeine for its stimulatory effects on the central nervous system. The usual dose is 200 mg contained in one or two small tablets, depending on their strength. It is also possible to get the same effect by drinking a caffeine-containing beverage, such as coffee or tea. Because coffee normally contains between 1 and 2 percent of the active constituent, it is necessary to extract up to 20 g (2/3 oz) of the product to yield that same amount. Tea contains more caffeine—up to 4 percent—but the method of preparation extracts less of it. Probably about 10 g (1/3 oz) of tea would be necessary to yield the same amount of caffeine found in one or two tablets. This assumes the beverage would be boiled during its preparation, rather than steeped, as is the usual custom.[1]

Dilution is not the only difference that must be considered in utilizing medicinal herbs. In addition to physiologically inert substances such as cellulose and starch, herbs often contain active principles that may be closely related both chemically and therapeutically to the constituent primarily responsible for its effects. Digitalis is a much-cited example of just such an herb. This leafy herb contains some thirty different closely related glycosides, all of which possess cardiotonic properties but which, due to small structural differences, have different speeds of onset of action and different durations of their effects. For example, one of the glycosides, digitoxin, when administered orally has an onset of action of one to four hours, with peak activity demonstrated at eight to fourteen hours. Another, digoxin, has an onset of action ranging from one-half to two hours and reaches a peak activity level in two to six hours.[2]

Certain digitalis plants contain both of these active principles along with many others. Proponents of its use argue that it is a very effective and useful drug because its multiplicity of constituents provides a uniform activity of short onset and long duration. Although this is true to some degree, it is also true that the activity of the leaf is difficult to standardize. The presently employed standardization procedure that involves cardiac arrest in pigeons is just one of a long series of biological assays utilizing such animals as frogs, cats, goldfish, chick hearts, and even daphnia (water fleas) in an attempt to obtain an accurate measure of potency. The ability to measure physiological potency in terms of the weight of a pure chemical entity is one of the principal reasons why administration of a purified constituent was considered advantageous in the first place. This concept is not new. It dates back to Paracelsus in the early sixteenth century.[3]

In addition to containing constituents with a desired activity, herbs often contain other principles that detract from their specific therapeutic utility. For example, cinchona bark contains some twenty-five related alkaloids, but the only one recognized as useful in the treatment of malaria is quinine. Thus, if powdered cinchona bark is administered as a treatment for malaria, the patient will also receive appreciable amounts of the alkaloid quinidine, a cardiac depressant, and cinchotannic acid, which, because of its astringent properties, would induce constipation.[4] Such side effects must be taken into account in the use of medicinal herbs.

Herbal quality

The matter of proper identification and appropriate quality—that is, lack of adulteration, sophistication, or substitution—is an extremely important one in the field of herbal medicine. Many of today's widely used herbs were once the subject of official monographs in *The United States Pharmacopeia* (USP) and *The National Formulary* (NF). These monographs established legal standards of identity and, subject to the limitations of the methods of the period, quality of the vegetable drugs.

No such standards exist today. Many of the herbs are collected in developing nations by persons who are not necessarily knowledgeable about the subtleties of plant taxonomy. They are usually sold to organizations that market them under their common names rather than the recognized Latin binomial; because of the lack of uniformity of the former, this can cause confusion. Likewise, the marketer may not employ the necessary skilled personnel or have the requisite analytical equipment to be able to establish identity and quality of the wide variety of botanicals sold. Herbs are sold in various ways, ranging from essentially whole plants or plant parts to cut pieces to finely ground powders. It is ordinarily impossible for the lay person to determine the quality or even the identity of the plant material by visual inspection. Because government standards of quality are nonexistent in the United States, the buyer is totally dependent upon the reputation of the seller. As a general rule, the larger firms have more to lose if they sell herbs of inferior quality, but some smaller organizations have outstanding reputations for marketing quality materials.

Returning to the matter of standardization of herbs or herbal extracts, it must be noted that the concentration of active constituents in different lots of supposedly identical plant material is highly variable. First of all, genetic variations exist. Just as one variety of apple tree will produce larger or tastier apples than another, so too will one variety of peppermint produce a larger quantity of or more flavorful peppermint oil than another, even though the conditions of growth remain identical. These genetic variations in medicinal herbs, many of which are obtained from plants growing in the wild, are not well understood.

Also of great importance to the quality of an herb are the environmental conditions under which it is grown. Fertility of the soil, length of growing season, temperature, amount of moisture, and time of harvest are some of the significant factors. Processing also plays a role. Some constituents are heat labile, and the plant material containing them needs to be dried at low temperatures. Other active principles are destroyed by enzymatic processes that continue for long periods of time if the herb is dried too slowly.

Because few of these factors are precisely controlled even for cultivated plants, let alone those harvested from the wild, the most effective way to ensure herbal quality is to assay—that is, to establish by some means—the amount of active constituents in the plant material. If the chemical identity of the constituent is known, it or a marker compound indicative of the activity of the herb can usually be isolated and quantified by appropriate physical or chemical methods. If it is unknown, if it is a complex mixture of constituents, or if no marker compound is available, biological assays such as that employed for digitalis must be utilized, at least initially. Once the potency of the herb is known, it can be mixed with appropriate quantities of material of greater or lesser potency to produce a product with defined activity. At the present time, standardization of the most widely used medicinal herbs is becoming quite common.

In terms of quality, the more expensive the plant material is, the more likely it is to be inferior. Finely powdered herbs or dosage forms such as tablets or capsules made from them are particularly susceptible to fraud by adulteration or substitution. One study of fifty-four ginseng products showed that 60 percent of those analyzed were worthless and 25 percent of them contained no ginseng at all.[5] A recent Canadian study showed that no North American feverfew product analyzed contained the recommended minimal content of 0.2 percent parthenolide believed to be required for effectiveness.[6] For many years, the root of prairie dock, *Parthenium integrifolium* L., was wrongly marketed as echinacea (*Echinacea* spp.)[7] and may still be in some isolated instances. Volatile-oil-containing botanicals often have their aromas enhanced by the addition of quantities of essential oils from other sources. This is said to be a common practice in the beverage herbal tea industry.

Although standardized extracts of ginkgo, ginseng, milk thistle, St. John's wort, and several other plants are available, the quality of unstandardized herbs in the American market today is extremely variable. This presents a problem because it makes the establishment of a specific dose difficult, if not impossible. It would be more of a problem if it were not for the fact that the therapeutic potency and potential toxicity of the active principles in many herbs are very modest; this, coupled with the great dilution in which they occur in the plant material, renders precise dosage often unnecessary.

It is also this lack of high therapeutic potency, accompanied with reduced side effects, that makes herbs more useful for the long-term treatment of mild or chronic complaints than for the rapid healing of acute illnesses. This is well illustrated in the current usage of feverfew as a preventive in cases of migraine or vascular headache. Although the herb has no utility in the treatment of acute attacks, accumulated evidence indicates its effectiveness in preventing such attacks if taken on a regular basis.[8]

Paraherbalism

Paraherbalism is faulty or inferior herbalism based on pseudoscience.[9] It is sometimes difficult to differentiate it from true or rational herbalism because its advocates use scientific and medical terminology, making it appear valid. However, on closer examination, one often finds that in paraherbalism medical claims are unsubstantiated by scientific evidence, the basis for the use of a particular herb may lack scientific logic, or clinical trials supporting use have been flawed (and sometimes combinations of these), thus providing questionable results. The ten tenets or precepts that distinguish paraherbalism from rational herbalism are worth presenting here in summarized form. Awareness of them will assist interested persons in distinguishing fact from fiction in a field where the former is scarce and the latter is abundant. Proceed with caution any time one of the following italicized statements appears in an herbal reference work or journal article:

1. *A conspiracy by the medical establishment discourages the use of herbs.*
   There is no conspiracy. Relatively few health care practitioners have any knowledge of the field. The pharmaceutical industry no longer views phytomedicinals as unprofitable products.

2. *Herbs cannot harm, only cure.*
   Some of the most toxic substances known—amatoxins, convallatoxin, aconitine, strychnine, abrin, ricin—are derived from plants.
3. *Whole herbs are more effective than their isolated active constituents.*
   For every example cited in support of this thesis, there is at least one example denying it. In addition, many herbs contain toxins in addition to useful principles. Comfrey is an example.
4. *"Natural" and "organic" herbs are superior to synthetic drugs.*
   Friedrich Wöhler, a German chemist, disproved the "natural" part of this in 1828 when he synthesized urea from inorganic starting materials. Yet we still find persons in this last decade of the twentieth century insisting that vitamin C from natural sources is in some way superior to vitamin C prepared synthetically from glucose. Established limits on pesticide residues probably render treated plants no more harmful than "organic" plants.
5. *The "doctrine of signatures" is meaningful.*
   This ancient belief postulates that the form of a plant part determines its therapeutic virtue. If it were true, kidney beans should cure all types of renal disease and walnuts various types of cerebral malfunction.
6. *Reducing the dose of a medicine increases its therapeutic activity.*
   There is no proof that this is universally true as espoused by practitioners of homeopathy. Positive results obtained by homeopathic treatment are either demonstrations of the placebo effect or of observer bias.
7. *Astrological influences are significant.*
   No scientific evidence supports this assertion.
8. *Physiological tests in animals are not applicable to human beings.*
   Differences do exist, but there is a high probability of significance and applicability when diverse animal species, especially those from different orders, show similar effects.
9. *Anecdotal evidence is highly significant.*
   It is extremely difficult to assess the reliability of such evidence. Consequently, it must be viewed simply as one of many factors (animal tests, clinical trials, etc.) that may tend to indicate the therapeutic utility of an herb.
10. *Herbs were created by God specifically to cure disease.*
    This thesis is not testable and should not be used as a substitute for scientific evidence.

Unfortunately, the practitioners of paraherbalism who espouse one or more of these tenets have done much to discredit the legitimate use of herbs for therapeutic purposes. For whatever reason— misguided belief, personal gain, or simply ignorance—they have flooded the market with literature containing so much outdated and downright inaccurate information about the use of herbs that interested individuals, lay or professional, who approach the field for the first time become totally confused. Because it serves their purposes, paraherbalists often accept at face value the disproved positive statements of a Renaissance herbalist such as Nicholas Culpeper[10] or a folk writer such as Maria Treben.[11] However, they discount the findings of modern science that demonstrate the toxicity of an herbal product.[12]

Paraherbalism has helped perpetuate the erroneous concept held by many in the medical community that herbal remedies have little or no pharmacological effect and, at best, are mere placebos that might do little good but do not cause any harm. Indeed, many physicians associate herbal medicine with worthless nostrums such as the snake oils and patent medicines promoted by the traveling medicine shows of the nineteenth century, part of a very colorful chapter in American history. On the other hand, some physicians believe that the use of herbs is actively dangerous, that claims made for them are outrageous, that harm is possible from self-medicating with them, and that practitioners of herbal medicine use empirical rather than scientific methods in evaluating their efficacy and safety.[13]

Homeopathy

Homeopathy is a particularly pernicious form of paraherbalism. In the late eighteenth century Samuel Hahnemann, a German physician, chemist, and pharmacist, formulated its basic principles.[14,15] He believed that symptoms of disease were an outward reflection of an imbalance in the body, rather than a direct manifestation of the illness itself. Treatment should therefore reinforce these symptoms, and the medicine used should produce similar symptoms in healthy individuals. This is the "law of similars" and is a fundamental tenet of homeopathy; consequently, the name is derived from the Greek *homoios,* meaning similar, and *pathos,* meaning suffering. Hahnemann and his followers conducted "provings" in which they administered herbs, minerals, and other medicinal substances to healthy people and assessed the symptoms that were produced. These detailed records were compiled into reference books used to match a patient's symptoms with a corresponding drug. The second and most controversial basic tenet of homeopathy is the "law of infinitesimals," which decrees that the potency of a drug is inversely proportional to its concentration.

In the preparation of homeopathic products, a wide variety of natural substances, predominantly herbs, is used to produce mother tinctures or powders. If the herbal constituents are soluble in water or alcohol, the herb is macerated in the liquid to prepare a tincture; if they are insoluble, the herb is finely ground and pulverized with powdered lactose. The mother tinctures and powders are then used to prepare the medicinal agent by serial dilution either 1:10, the decimal system designated 1X, or 1:100, the centesimal system designated 1C. After each dilution, the liquid preparations are vigorously shaken and tapped on a resilient surface, a process known as "succussion," and solid preparations are vigorously pulverized in a mortar and pestle. Most remedies range from 6X (one part per million) to 30X, but products of 30C or more are marketed.

It is interesting to note that a 30X dilution means that the original substance has been diluted 10^{30} times; however, this dilution would exceed Avogadro's number (6.02×10^{23} molecules are present in 1 g molecular weight of a substance), and not a single molecule of the drug would be present. Yet cures are claimed for dilutions as high as 2000X. The belief is that the vigorous shaking, tapping, or pulverizing with each step of dilution transfers the inherent essence, or "spirit," of the drug to the inert molecules of the solvent or lactose. This mystical phenomenon is completely contrary to everything that is known about the scientific laws that operate in the physical world.

With only about 300 licensed practitioners—half of them physicians and the rest mostly chiropractors, naturopaths, dentists, veterinarians, or nurses—homeopathy is not widely practiced in the United States. Larger numbers of homeopaths practice in Europe and India. In the United States, the greatest problem arising from homeopathy is that homeopathic remedies have tainted the medical and scientific communities' perception of herbal medicine. Because the *Homeopathic Pharmacopeia of the United States* was developed in the nineteenth century, the majority of the drug entries are herbal medicines; consequently, a major portion of the homeopathic remedies available today from practitioners, pharmacies, and health food stores is based on nineteenth-century materia medica—namely, herbal drugs. Unfortunately, these ineffective homeopathic nostrums have been equated with all herbal medicines in general. This is erroneous, and it is important to remember that herbal medicines other than homeopathic preparations may contain potent bioactive chemicals that could be therapeutically efficacious or could be hazardous to one's health if not used properly.

The problem is compounded because homeopathic remedies constitute the only category of bogus products legally marketed as drugs in the United States. They have gained this unique status because most of these remedies were on the market before the passage of the Food, Drug, and Cosmetic Act of 1938. Due to the efforts of U.S. Senator Royal Copeland, who was also a homeopathic physician, homeopathic remedies were exempted from the law that required drugs to be proven safe.

In 1962, the law was amended to require drug manufacturers to prove that their products are also effective for treating indicated conditions. Unfortunately, the amendments did not rescind the 1938 exemption for homeopathic drugs. Recently, a petition by physicians, scientists, and consumer advocates was submitted to the U.S. Food and Drug Administration (FDA) calling for that agency to stop the marketing of OTC homeopathic products until they have been shown to meet the same standards of safety and effectiveness as other OTC drugs. The FDA has not acted on this petition, so in the meantime, *it is recommended that pharmacies not stock homeopathic products and that patients be warned that these products are not proven efficacious.*[16]

Rational herbalism

Paraherbalism and homeopathy have done much to discredit herbal medicine, and those seeking to establish its validity start with a severe handicap. Nevertheless, a knowledge of historical drug development will indicate immediately that plants have long served as a useful and rational source of therapeutic agents. Not only do plant drugs such as digitalis, the opium poppy, ergot, cinchona bark, psyllium seed, cascara sagrada, rauwolfia, belladonna, and coca leaves continue to serve as useful sources of pharmaceuticals, but their constituents also serve as models for many of the synthetic drugs used in modern medicine.

With few exceptions, drug development has followed a logical progression from an unmodified natural product, usually extracted from an herb, to a synthetic modification of that natural chemical entity to a purely synthetic compound apparently showing little relationship to its natural forebears. An example of this type of development, illustrated in Figure 1.1, can be seen with the widely used drug ibuprofen. It was introduced to the American market in 1974 as one of the first of a new generation of nonsteroidal anti-inflammatory drugs (NSAIDs). These drugs possess chemical structure features similar to those found in aspirin (acetylsalicylic acid) and were developed to reduce undesirable side effects associated with prolonged aspirin usage such as increased bleeding tendencies and gastrointestinal irritation.

However, the original NSAID was not aspirin but, rather, salicin, which was first isolated in 1829 from the bark of the willow tree (*Salix* spp.) by the French pharmacist H. Leroux. The first modern reference to the use of willow bark was in treating the fever of malaria and dates from 1763, although the virtues of the plant were known by the ancients, including Hippocrates. Salicin was shown to be a prodrug that was converted to salicylic acid in the intestinal tract and liver after oral administration. Salicylic acid, when taken orally, possesses excellent anti-inflammatory, analgesic, and antipyretic properties; however, even its sodium salt is difficult to tolerate for lengthy periods because of the irritation and damage it produces in the mouth, esophagus, and, particularly, the stomach. Aspirin was introduced into medicine in 1899 and found to be equally effective and better tolerated.[18]

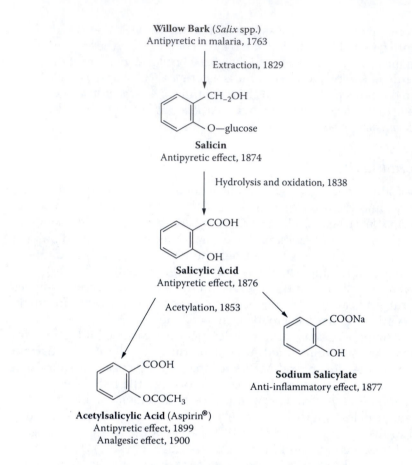

Figure 1.1 Chronology of drug development from willow bark.

At the present time the production of NSAIDs for the treatment of pain, fever, and inflammation represents sales worth billions of dollars per year for the drug industry in the United States alone. It is amazing when one considers that this enormous NSAID production all grew from the common willow tree. It is just one of numerous examples that show how much the modern pharmaceutical industry owes to its natural-product heritage.

It is not unreasonable to expect that, of the thirteen thousand plant species known to have been used as drugs throughout the world (some of them for centuries), there are still many with useful therapeutic properties that have been little studied. Thus, although we may as yet be unable to isolate, purify, and market a botanical's chemical constituent for use as a drug, the herb that contains it may still be used in its natural or phytomedicinal form, and desirable therapeutic effects may be achieved.

It is, therefore, important to realize that, although herbs are literally diluted drugs, the nature of the active principles in them is often a matter of empirical observation and tradition rather than the result of extensive clinical testing. The reasons for this lack of clinical testing are basically economic because the cost of such evaluation is extremely high.

However, perhaps even more important than whether an herbal remedy works (i.e., has the desired therapeutic utility) is the matter of whether it is safe. It might be surmised that herbs consumed by humans for generations, centuries, and even millennia must be reasonably safe. This has generally proven true, at least insofar as acute toxicity is concerned. But it is not necessarily true for some of the newly introduced, more exotic products that are continually being placed on the market by herbal enthusiasts. Nor is it the case for some of the older herbs that have recently been shown to produce deleterious effects of a chronic, more subtle nature following long usage. Certain comfrey species, with their content of toxic pyrrolizidine alkaloids, are an excellent example of this latter kind of herb.[19]

Some have questioned the validity of herbal medicine because they doubt the wisdom of self-treatment for diseases of any kind, and, as previously noted, most herbal treatments in the United States are self-selected. That this point of view is largely invalid is obvious to anyone who has followed the recent trend in OTC drugs. A number of significant drugs, including hydrocortisone, ibuprofen, clotrimazole, and various antihistamines, have been converted from prescription to OTC status, and this trend is continuing.

Many books dealing with "unconventional" or "alternative" medicine contain, along with discourses on such subjects as acupuncture and homeopathy, a chapter or two on herbal medicine. This shows a complete lack of understanding on the part of the authors of such reference works. *Rational herbal medicine is conventional medicine.* It is merely the application of diluted drugs to the prevention and cure of disease. The fact that the constituents and, sometimes, even the mode of action of these drugs are often incompletely understood and that instruction in their appropriate application is not a significant part of standard medical curricula does not in any way detract from their role in conventional medicine. If it did, we would be forced to discontinue the use of a number of popular products such as psyllium seed and senna laxatives, together with about 25 percent of our current materia medica that is derived from such sources.

We would also have to conclude that some 62 percent of all German adults are unconventional because that is the percentage that turns first to a natural remedy (herbs or phytomedicines) to treat their illnesses.[20] Chinese medicine has become a hybrid of Western and traditional practices, but personal observation indicates that if the use of herbal medicine were considered unconventional, almost the entire population of China would fall into that category. Although herbal therapy may not be mainstream American medicine, it certainly is conventional.

General guidelines in the use of herbal medicines

With respect to self-medication with herbal medicines, it is important to know the conditions for which one can treat oneself and those that deserve professional medical care. The occasional pain of a headache or a strained muscle, a mild digestive upset or simple diarrhea, infrequent insomnia, the common cold—all are conditions that are amenable to and usually receive self-treatment. On the other hand, rheumatoid arthritis, cardiac arrhythmia, or cancer requires professional medical care. Self-treatment in such conditions would be utter folly.

Generally speaking, consumers lack the background to diagnose most clinical conditions accurately. Because one disease can mimic another, potentially serious conditions can be misdiagnosed. Effective self-care requires highly informative and understandable package labeling and patient education materials that emphasize safe, appropriate, and effective use. Unfortunately, in the United States, the FDA does not regulate herbal medicines as drugs but rather as dietary supplements; consequently, health or therapeutic claims cannot be placed on the package label. However, a vast hyperbolic advocacy literature has built up around them, providing product information designed to promote sales, not necessarily to inform. This situation increases the need for health care professionals, especially pharmacists, to judge the quality of available products and to interpret the products' role in preventing and treating disease for the lay consumer.[21,22]

The FDA neither establishes nor regularly enforces any standards of quality for herbal products. This means that one must rely upon the reputation of the producer for any quality assurance. Products are often misbranded, and often the quantities of the ingredients are not listed. In mixtures containing large numbers of herbal constituents, quantities sufficient to render a therapeutic effect may be lacking. The consumer is best advised to purchase a preparation containing a specified amount of a standardized extract marketed by a reputable firm. Some herbal producers have responded to the question of quality by introducing entire lines of products, all standardized so that each unit of dosage contains a specific quantity of active constituents. Sales of some of these product lines are restricted by manufacturers to pharmacies in which patients can receive professional advice. In addition to standardization of the herbal preparation, another indicator of quality assurance is that the label shows the scientific name of the botanical, the name and address of the actual manufacturer, a batch or lot number, the date of manufacture, and the expiration date.

Technically, most herbal medicines are unapproved drugs. They may have been used for centuries, but substantial data on the effectiveness and safety of long-term use are often lacking. Safety considerations include warnings and precautions relative to the use of a particular herb, drug–drug interactions between prescription medications and the herbal medication, and the fact that certain groups of individuals often experience a higher incidence of adverse drug effects that could have dire consequences. In the case of warnings and precautions, patients should cease taking an herb immediately if adverse effects (allergy, stomach upset, skin rash, headache) occur.

Another important safety precaution is that herbal use is not recommended for pregnant women, lactating mothers, infants, or children under the age of six. In the pregnant woman, most drugs cross the placental barrier to some extent, and these expose the developing fetus to potential teratogenic effects of the drug. The first trimester, when organogenesis occurs, is the period of greatest teratogenic susceptibility and is the critical period for inducing major anatomical malformations. In the lactating mother, the potential exists for the secretion of the drug in the mother's milk, resulting in adverse effects in the nursing infant. The body and organ functions of infants and young children are in a continuous state of development. Changes in the relative body composition (lipid content, protein binding, and body-water compartments) will produce a different drug distribution in their bodies than in adults. Also, many enzyme systems may not be fully developed in infants, particularly neonates, thus producing a slower drug metabolism.

The elderly patient should be particularly cautious in using herbal medicines. It is well documented that the aging process results in a significant increase in the proportion of body fat to muscle mass and in decreases in renal function, total body water, lean body mass, organ perfusion, and hepatic microsomal enzyme activity. Such changes may lead to altered absorption, distribution, metabolism, and elimination of drugs that, in turn, could result in an accumulation of a drug in the body all the way to toxic levels.

Therefore, with the elderly patient, careful attention should be given to drug dosing and monitoring for adverse effects. In addition, because elderly patients may tend to have serious and multiple diseases for which they are taking prescription medications, drug–herb interactions may be a major concern. For this reason—not only for the elderly, but also for all other patients—a patient's drug history should always include herbal medicines taken, and individuals should be advised to share information on self-medication with herbal medicines with their physician and pharmacist.[23]

Herbal dosage forms

Herbs are consumed in various ways, most commonly in the form of a tea or tisane prepared from the dried plant material. Both of these terms refer to what is technically known as an infusion, prepared by pouring boiling water over the herb and allowing it to steep for a period of time. In such cases, the herb and the water are not boiled together. Time of steeping is important, for many of the desired components are not very water soluble. For example, in the case of chamomile, where much of the desired activity is present in the volatile oil, even a prolonged steeping of ten minutes extracts only about 10–15 percent of the desired components.[24]

Occasionally, an herbal preparation is made by boiling the plant material in water for a period of time, then straining and drinking the resulting extract. Technically, the process is called decocting, and the resulting liquid is known as a decoction. Boiled coffee is prepared in this way as opposed to beverage tea, which is an infusion.

The quantity of herb to be extracted is usually rather imprecise, being stated in terms of level or heaping teaspoonfuls. A standard teaspoonful of water weighs about 5 g (approximately 1/6 oz) and a heaping teaspoonful of most herbal materials approximately half that (2.5 g). Very light herbs such as the flower heads of chamomile may weigh only 1.0 g per heaping teaspoonful. The same quantity of a leaf drug might weigh 1.5 g and of a root or bark about 4.5 g. Even these weights are variable according to the degree of comminution (chopping or grinding) of the plant material. Finely powdered herbs are obviously going to have less space among the particles, and an equal volume will weigh more. A heaping teaspoonful of finely powdered ginger weighs about 5 g. Standard instructions for the preparation of a tea call for one heaping teaspoonful per cup of water (240 mL or 8 oz). It must be noted that the stated sizes of teaspoons are those established by long-standing convention. Experience indicates that modern teaspoons have a capacity some 25 percent greater than these standards.[25] For most herbal preparations, this difference is probably of minor significance.

As previously noted, herbs are often marketed as phytomedicinals in various dosage forms. Some of the more common ones include powders in hard gelatin capsules or, together with suitable fillers and binders, as compressed tablets. When the active principles are not soluble in water or when a more concentrated product is required to allow adequate dosage, various extracts of herbs are prepared. These are usually hydroalcoholic solutions or tinctures of such concentration that 10 mL will contain the active constituents in either 1 or 2 g of herb. Even more concentrated preparations are known as fluidextracts, 1 mL of which represents 1 g of plant material. Tinctures or fluidextracts are consumed as such or by diluting a specific quantity—usually a certain number of drops—in water so that it may be easily swallowed. The most concentrated form of an herb is a solid extract prepared by evaporating all of the solvent used to remove the active constituents from the herb. Extracts are often available in powdered form; 1 g usually represents 2–8 g of the starting material. They are normally encapsulated for ease in administration.

If consumers decide to purchase the herbs themselves, rather than a processed dosage form, they should remember certain guidelines that will help ensure the acquisition of a quality product. Lacking expert knowledge, however, there is no sure way to avoid all of the pitfalls. Buying clean, dried herbs that are as fresh as possible from a reputable source—preferably in a form allowing positive identification—and ensuring freedom from insect infestation by careful inspection is probably the best general advice. Many herbs are valued for their aromatic principles. These are stable in whole plant parts for a longer period of time than in finely powdered material. Such plant parts can easily be reduced to the desired fineness in a small electric coffee grinder just before making a tea. Maintain all herbs in tightly closed, preferably glass containers, away from sunlight and in a cool, dry place.

Herbal medicine information sources

Essential to any type of effective health care is the availability of objective, unbiased information based on truth and accuracy. Unfortunately, in the case of herbal medicine, many information sources are inaccurate and sensationalize or distort the information they contain. In addition, obtaining reliable information on herbs has become a daunting task because there is an herbal medicine renaissance; information sources abound, including, among many others, news magazines and television, books on paraherbalism and herbalism, scientific and professional journals, and online databases.[26]

In the evaluation of the safety and efficacy of drugs, the gold standard is the controlled clinical trial based on scientific methodology in which bias has been eliminated. In a controlled trial, patients in one group receive the investigational drug. Those in the control group usually get either a placebo, a different drug known to be effective, or a different dose of the drug under study. Patients are randomly assigned to either the treatment or the control group in order to eliminate selection bias. In conjunction with randomization, an experimental design feature known as "blinding" helps ensure that bias does not distort the results of the study. In a double-blind study, neither the patients nor the investigators know which patients received the investigational drug. Only when the secret assignment code is broken do the individuals in the study know which patients were in the treatment or control groups.[27]

Information on the results of clinical trials is published in peer-reviewed scientific and professional journals. The peer-review process means that before an article is accepted for publication, it is scrutinized by scientists and professionals working in the same area in order to ensure that the conclusions derived from the results of the study are valid. In addition, peer-reviewed professional journals publish lengthy, detailed observations made by professionals about patients in their care. These patient case histories, although lacking the control-study aspect, provide valuable suggestions of drug effects and potential safety problems.

Books constitute another important and widely used source of information on herbal medicine. The most valuable volumes are those that evaluate the current scientific and professional literature and that supply references to this primary literature. Unfortunately, the majority of the books available fall into the category of paraherbalism. These books run the gamut from glossy, expensive picture books with little accurate therapeutic information to books written by herbal medicine practitioners who discuss herbs from the viewpoint of their personal bias rather than on the basis of scientific fact. In general, an important caveat when using any information source is to be particularly cautious if references to the primary literature are lacking. Without these citations, there is no way to check the accuracy of the facts presented.

Finally, a particularly valuable information source on the safety and efficacy of herbs is the German Commission E monographs. They were used extensively in writing this book and are discussed in Chapter 3. Formerly, the information in the monographs was not readily available to those who did not read German; however, a complete English translation is now available.[28]

These, then, are the basic principles of herbal medicine as they apply to its current practice in the United States. The field is a curious mixture of ancient tradition applied to modern conditions without, in many cases, the benefit of modern science and technology. To be totally effective, the traditional practices must eventually be coupled with up-to-date scientific methodology. The reason that has not yet been done, except in certain isolated circumstances, will become clear in the Appendix, which examines the present laws and regulations pertaining to herbs in the United States.

The cooperation between empiricism and science—so urgently needed to bring herbs and phytomedicines into the mainstream of modern medicine—is a subject that has been explored in some depth by B. Lehmann, a German physician. Her conclusion, given here in translation, is especially pertinent and serves as a fitting summary to this chapter on basic principles[29]:

Phytomedicines, exactly like other medicines, must stand up to the challenge of modern scientific evaluation. They need no special consideration when it comes to the planning and conduct of clinical trials intended to prove their safety and efficacy. The distinctive feature of phytotherapy is its origin, namely, the many years of empirical use of plant drugs. Experience gained during this period should be taken into account, along with clinical testing, in evaluating the effectiveness of phytomedicines.

In 2005, a Swiss–British study evaluated 110 placebo-controlled homeopathy trials and 110 matched conventional-medicine (allopathy) trials. The authors concluded that the effects seen in placebo-controlled trials of homeopathy are compatible with the placebo hypothesis. In contrast, identical methods demonstrated that the benefits of conventional medicine are unlikely to be explained by unspecific effects.[17]

References

1. Tyler, V. E. 1987. *The new honest herbal, 53–56*. Philadelphia, PA: George F. Stickley.
2. Robbers, J. E., M. K. Speedie, and V. E. Tyler. 1996. *Pharmacognosy and pharmacobiotechnology, 117–119*. Baltimore, MD: Williams & Wilkins.
3. Sonnedecker, G. 1976. *Kremers and Urdang's history of pharmacy*, 4th ed., 40–42. Philadelphia, PA: J. B. Lippincott Company.
4. Robbers, J. E., M. K. Speedie, and V. E. Tyler. 1996. *Pharmacognosy and pharmacobiotechnology, 155–158*. Baltimore, MD: Williams & Wilkins.
5. Ziglar, W. 1979. *Whole Foods* 2 (4): 48–53.
6. Awang, D. V. C., B. A. Dawson, D. G. Kindack, C. W. Crompton, and S. Heptinstall. 1991. *Journal of Natural Products* 54:1516–1521.
7. Foster, S. 1991. *Echinacea: Nature's immune enhancer, 84–92*. Rochester, VT: Healing Arts Press.
8. Hobbs, C. 1989. *HerbalGram* 20:30–33.
9. Tyler, V. E. 1989. *Nutrition Forum* 6:41–44.
10. Culpeper, N. 1814. *Culpeper's complete herbal and English physician enlarged*. London: Richard Evans, 398 pp.
11. Treben, M. 1980. *Health through God's pharmacy*. Steyr, Austria: Wilhelm Ennsthaler, 88 pp.
12. Heinerman, J. 1979. *The science of herbal medicine*, xvi–xxi. Orem, UT: Bi-World Publishers.
13. Weil, A. 1989. *Whole Earth Review* 64:54–61.
14. Barrett, S., and V. E. Tyler. 1995. *American Journal of Health-System Pharmacy* 52:1004–1006.
15. Der Marderosian, A. H. 1996. *Journal of the American Pharmaceutical Association* NS36:317–328.
16. Skolnick, A. A. *Journal of the American Medical Association* 272:1154–1155.

17. Shang, A., K. Huwiler-Müntener, L. P. Nartey, S. Dörig, J. A. C. Sterne, D. Pewsner, and M. Egger. 2005. *Lancet* 366:726–732.
18. Robbers, J. E., M. K. Speedie, and V. E. Tyler. 1996. *Pharmacognosy and pharma-cobiotechnology,* 11–13. Baltimore, MD: Williams & Wilkins.
19. Awang, D. V. C. 1987. *Canadian Pharmaceutical Journal* 120:100–104.
20. *Welt am Sonntag,* March 23, 1997, p. 40.
21. Tyler, V. E. 1996. *Journal of the American Pharmaceutical Association* NS36:29–37.
22. Tyler, V. E., and S. Foster. 1996. Herbs and phytomedicinal products. In *Handbook of nonprescription drugs,* 11th ed., ed. T. R. Covington, 695–713. Washington, D.C.: American Pharmaceutical Association.
23. Klein-Schwartz, W., and B. J. Isetts. 1996. Patient assessment and consulta-tion. In *Handbook of nonprescription drugs,* 11th ed., ed. T. R. Covington, 11–20. Washington, D.C.: American Pharmaceutical Association. .
24. Robbers, J. E., M. K. Speedie, and V. E. Tyler. 1996. *Pharmacognosy and pharma-cobiotechnology,* 87. Baltimore, MD: Williams & Wilkins.
25. Reich, I., E. T. Sugita, and R. L. Schnaare. 1995. Metrology and calculation. In *Remington: The science and practice of pharmacy,* 19th ed., ed. A. R. Gennaro, 63–73. Easton, PA: Mack Publishing Company.
26. Hoffman, E. 1994. *The information sourcebook of herbal medicine,* 1–60. Freedom, CA: The Crossing Press.
27. Flieger, K. 1995. Testing drugs in people. In *From test tube to patient: New drug development in the United States,* 6–11. FDA Consumer Special Report, DHHS pub. no. FDA 95–3168. Rockville, MD: Department of Health and Human Services.
28. Blumenthal, M., W. R. Busse, A. Goldberg, T. Hall, C. W. Riggins, and R. S. Rister, eds. 1998. *The complete German Commission E monographs: Therapeutic guide to herbal medicines,* trans. S. Klein and R. S. Rister. Austin, TX: American Botanical Council.
29. Lehmann, B. 1992. *Zeitschrift für Phytotherapie* 13:14–18.

chapter two

Contents and use of subsequent chapters

The herbal monographs found in subsequent chapters do not constitute a comprehensive, encyclopedic listing of all of the more than three hundred herbs currently used in Western medicine. Readers who anticipate that approach will not have their expectations realized. The chapters that follow are instead devoted principally to phytomedicinals that are now considered to be the most useful for treating particular diseases or syndromes. Following brief general discussions of the pathophysiology of the various conditions, monographs of the useful herbs for treating those disorders are arranged in approximate order of their decreasing therapeutic utility. Occasionally, a particular section will contain brief discussions of herbs that are not particularly effective but are nevertheless included because of their popularity. Several minor carminatives in Chapter 3 are a case in point. A very few herbs considered totally ineffective or even dangerous to use are included for the same reason. Sarsaparilla and sassafras in Chapter 11 are examples.

Simply because an herb is listed does not mean that it should be used for the particular ailment. For example, phytomedicinals have no place in the self-treatment of self-diagnosed heart disease or cancer. Herbs of potential value in such conditions are discussed to bring them to the attention of professionals who are qualified to use them properly. Hawthorn and taxol are examples. Each monograph must be read carefully to determine the safety, utility, and proper use of the herb discussed there.

Unlike the botanical or alphabetical system utilized in most herbals, the classification of phytomedicines in this volume is based on their principal therapeutic use. Because some of them are useful for more than one condition, they may appear more than once. In most such cases, the minor reference is cross-referenced to the major monograph. For example, chamomile preparations are quite useful in the treatment of various kinds of skin conditions. However, such pharmaceutical preparations are not ordinarily available in the United States, so chamomile is briefly mentioned under treatments for dermatitis in Chapter 10, and the reader is referred to Chapter 3. There the herb is discussed in detail, principally as a digestive aid, but also with respect to its other beneficial properties.

The contents of the major monographs follow the same general pattern with minor deviations. Ordinarily, the part of the plant used, the scientific name (Latin binomial followed by author citation), and the plant family are presented first, but synonyms, unless they are especially meaningful, are ignored. In a work devoted primarily to the therapeutics of useful herbs, enumeration of the multiplicity of common names was deemed unnecessary, as was much pharmacognostical information, such as habitat, production, preservation, and marketing. All of this is readily available elsewhere.

Chemical identification of the active principles (when known) is followed by a discussion of therapeutic use, mechanism of action, side effects, dosage and dosage forms, and usually some remarks about the value of the herb in the eyes of an authority such as the U.S. Food and Drug Administration (FDA) or Commission E of the former German *Bundesgesundheitsamt* (federal health agency), with additional comments by the authors. The FDA and its attitude toward phytomedicinals have been discussed sufficiently in Chapter 1. The activities of Commission E are probably less well known to American readers, but because they are frequently quoted in the monographs, some explanation of that commission's role in the evaluation of the safety and efficacy of phytomedicinals is required.

In 1978, the German equivalent of our FDA, formally known as the *Bundesgesundheitsamt,* undertook the task of evaluating the safety and efficacy of phytomedicinals. To do this, the agency established Commission E, which was presented with the formidable task of examining appropriate data concerning about fourteen hundred different herbal drugs corresponding to some six hundred to seven hundred different plant species. The data utilized by the commission included results obtained from clinical trials, collections of single cases, and scientifically documented medical experience. The latter category comprises both the scientific literature and collective conclusions of medical associations.[1]

Results of the study were originally published in German as Commission E monographs in the *Bundesanzeiger* (*Federal Gazette*) and are now available as a single volume in the English translation.[2] More than three hundred such monographs, covering most of the economically important herbal remedies sold in Germany, appeared by 1993.[3] Approximately two-thirds of the monographs provide positive assessments of herbs found to be safe and effective. The remaining monographs are negative, usually because the herb presents an unsatisfactory risk–benefit ratio. Each monograph generally provides summaries of such information as the identity, composition, use, contraindications, side effects, precautions, dosage, preparations, and effects for those herbs considered effective. For ineffective herbs, comments on the risks involved in consumption and an overall evaluation are substituted for much of the preceding information.

Of the first 285 monographs published, 66 percent mention risk aspects; of these, fifty-eight monographs state that there is no plausible evidence of efficacy, and thus a negative risk–benefit ratio exists. Concerning contraindications, sixty-three monographs mention allergy to the active constituent, twenty-four restrict use during pregnancy or lactation, fifteen state contraindication if the patient has gallstones, and seven monographs are contraindicative if the patient suffers from an inflammatory disease of the kidney. Common types of side effects are gastrointestinal disorders (mentioned in thirty-five monographs), allergic reactions (recorded in thirty), and photosensitivity (identified in five).[4] See Blumenthal et al. for additional details.[2]

The Commission E monographs contain the best evaluations of the utility of phytomedicinals currently available. Most experts in the field are in general agreement with the findings reported therein. Occasionally, a different opinion has been expressed. If this is the case, in this volume the Commission E judgment is tempered somewhat by calling the herb "apparently modestly effective in this regard." An example of a questionable statement in a Commission E report is the therapeutically noneffective dose of salicin (60–120 mg) recommended in the monograph on willow bark. Because of some scattered controversial findings of this nature, it is most unfortunate that neither the data used by the Commission in reaching its conclusions nor the minutes of its meetings are available to the public.

Nevertheless, to repeat for emphasis, the findings of the Commission E on herb safety and efficacy constitute the most accurate body of scientific knowledge on that subject available in the world today. They are extensively referred to in this book.

Numerous references are included in the discussions of the various herbs in the following chapters. These will enable any of the principal subjects presented therein to be pursued in additional detail. Although it is regretted that some of the key references are written in the German language, this simply reflects the country in which many of the investigations have been carried out. The discussion of the various properties of carminatives is an example where comparable information is not available in English. A considerable amount of information on herbal use appears for the first time in English in this book.

Some readers may consider it unusual that herbs commonly utilized in traditional Chinese and Indian (Ayurvedic) medicine are not represented more frequently in the subsequent chapters. Many possible entries were considered, but with rare exceptions, such as ephedra, their utility remains unproven by Western scientific standards. The philosophic principles of Chinese and Ayurvedic medicine differ vastly from those on which Western medicine is based and lead, in some cases, to very different conclusions. For example, in this work, Chinese ginseng and American ginseng—different species of the same genus that contain similar active principles—are considered together. As far as can be determined by Western studies, they have similar actions. Yet the Chinese believe they are very different in their properties: The American species is "cold" (yin) and the Oriental is "hot" (yang). In addition, many pharmacological and clinical studies carried out in China and India have lacked adequate "blinding" and controls in their experimental designs; their results are thus very difficult to interpret.

For these reasons, most of the data on which the herbal information in this book is based has been derived from studies in Europe and America. Because of their vast floras, China and India will eventually enrich herbal medicine with remedies of proven value; however, to date, their contributions have been minimal.

Every effort has been made to be accurate and fair in the presentation of the truly useful herbal remedies. Occasionally, the potential use of a product has been mentioned, but on the whole, the scope of the information has been restricted to statements that are presently substantiated by facts. But in both science and medicine, today's facts may not be the same as those of tomorrow. Therefore, as befits lifelong health professionals, the authors have attempted to present the information in neither a conservative nor a liberal manner but, rather, in a conscientious one. We believe that most readers will appreciate this concept as they peruse the pages that follow.

References

1. Keller, K. 1991. *Journal of Ethnopharmacology* 32:225–229.
2. Blumenthal, M., W. R. Busse, A. Goldberg, T. Hall, C. W. Riggins, and R. S. Rister, eds. 1998. *The complete German Commission E monographs: Therapeutic guide to herbal medicines*, trans. S. Klein and R. S. Rister. Austin, TX: American Botanical Council.
3. Schilcher, H. 1993. *Zeitschrift für Phytotherapie* 14:132–139.
4. Keller, K. 1992. *Zeitschrift für Phytotherapie* 13:116–120.

chapter three

Digestive system problems

Nausea and vomiting (motion sickness)

Ordinarily preceded by feelings of discomfort and uneasiness, nausea and vomiting are the principal results of motion sickness. These events are controlled by an emetic center in the brain that is influenced by stimuli from peripheral sites, from a so-called chemoreceptor trigger zone (CTZ) in the brain, or from the cortex of that organ. Antiemetics usually function by blocking these stimuli. Antihistamines—dimenhydrinate, for example—prevent peripheral stimuli from reaching the emetic center and are therefore particularly useful in cases of motion sickness or inner ear dysfunction. Anticholinergics, such as scopolamine, are also effective.

However, both of these useful types of drugs are not without unpleasant side effects. Many antihistamines produce drowsiness and subsequently impair mental and physical abilities; the anticholinergics may cause effects ranging from dry mouth and drowsiness to blurred vision and tachycardia. Effective medicines lacking such side effects are considered highly desirable. Although lacking central nervous system (CNS) effects and functioning by a mechanism not well understood, the herbal remedy of choice for preventing motion sickness is ginger.

Herbal remedy for nausea and vomiting

Ginger

Often mislabeled as a root, ginger is technically a rhizome or underground stem of the plant *Zingiber officinale* Roscoe. Commercial varieties are ordinarily designated according to their geographical origin, such as African ginger, Cochin ginger, or Jamaican ginger. Used in China as a spice and a drug for some twenty-five centuries, it is now employed throughout the world for both these purposes, but particularly as a flavoring agent. Its uses in traditional medicine are numerous, but the properties for which it has been most valued are those of a carminative or digestive aid.[1]

Results of the first clinical investigation into the antinausea and antiemetic potential of ginger were published in 1982.[2] Mowrey and Clayson tested thirty-six college students with a high susceptibility to motion sickness and concluded that 940 mg of powdered ginger was superior to 100 mg of dimenhydrinate in reducing the symptoms of motion sickness when consumed twenty to twenty-five minutes prior to tests conducted in a tilted rotating chair. The placebo chosen, curiously, was chickweed (*Stellaria media*).

The results of investigations subsequent to the initial one of Mowrey and Clayson have been quite interesting. An American group consisting of Stott and colleagues reported, in 1985, that ginger had no effect on motion sickness.[3] In 1986, Swedish investigators Grøntved and Hentzer found ginger significantly better than a placebo in reducing vertigo; it did not have any effect, however, on the nystagmus (oscillation of the eyeballs) of test subjects.[4] Then, Americans led by Wood reported negative results in 1988.[5] They noted that motion sickness is a CNS reaction to vestibular (inner ear) stimulation with only secondary gastrointestinal involvement. Because ginger lacks CNS activity, it was not, they concluded, a useful motion sickness preventive.

More recent European studies have yielded favorable results. Grøntved and colleagues tested the effects of ginger on seasickness in eighty Swedish naval cadets.[6] They concluded that ginger reduced vomiting and cold sweating. Also, fewer symptoms of nausea and vertigo were reported, but there is a question about the statistical significance of these observations. Schmid and others[7] compared seven commonly used agents for prophylaxis of seasickness, including ginger (Zintona™, Dalidar Pharma, Israel), in the form of an extract of the rhizome, standardized to pungent phenolics; similar rates of efficacy were observed for ginger and five of the six drugs tested. Only the Scopolamine transdermal therapeutic system showed a slight trend to less efficacy and more illness, causing slightly more visual problems than the other study medications.

In a randomized, double-blind trial, Riebenfeld and Borzone assessed the comparative efficacy and tolerability of Zintona and dimenhydrinate in sixty subjects (ages ten to seventy-seven) with a history of seasickness sensitivity.[8] Patients were administered either 500 mg of the ginger preparation thirty minutes before embarkation, followed by 500 mg every succeeding four hours, or 100 mg dimenhydrinate thirty minutes before embarkation, followed by 100 mg every succeeding four hours, both over the forty-eight-hour duration of the treatment. Ginger was judged very effective ($n = 21$) or effective ($n = 7$) in most cases as compared to very effective ($n = 15$) and effective ($n = 12$) for dimenhydrinate.

In Germany, Holtmann and colleagues carried out tests to determine the mechanism of ginger's positive effects.[9] They concluded that the drug does not influence the inner ear or the oculomotor system, both of which are of primary importance in motion sickness. Thus, a CNS effect of ginger should be ruled out, and it must be concluded that the antiemetic mechanism of action of ginger is of a gastrointestinal nature.

Concerning the involvement of the gastrointestinal tract, the symptoms of motion sickness are associated with an abnormal increase in pacemaker activity in the gastric musculature, called tachygastria, resulting in disordered gastric mobility. In addition, a consequence of this gastric dysrhythmia is a prolonged inhibition of gastric emptying. Therefore, drugs that prevent tachygastria or enhance gastric emptying may reduce the incidence and severity of motion sickness. Stewart and coworkers at Louisiana State University investigated the effects of ginger on motion sickness susceptibility and gastric function; however, even though they found that ginger partially inhibited tachygastria, it did not affect gastric emptying and did little to relieve the symptoms of motion sickness.[10] It may be relevant to note that this study was neither double blind nor random.

Two systematic reviews of randomized controlled trials (RCTs) of ginger for the prevention of nausea and vomiting have been conducted since publication of the 2nd edition of this volume. Ernst and Pittler[11] found six studies worthy of consideration; highest ranked were Grontved and others[6] and Arfeen and others[12], results of the former positive and the latter negative. Betz and others[13] discerned as potentially relevant 24 RCTs, of 100 published reports considered; of these 24 RCTs covering 1073 patients, 16 contained information regarding activity of ginger against kinetosis, postoperative nausea and vomiting (PONV), morning sickness, and hyperemesis gravidarum. The authors of the latter study concluded that while there was no clear evidence for the efficacy of ginger in the treatment of PONV and kinetosis, the results for the treatment of nausea and vomiting in pregnancy are encouraging, and in doses of up to 6 g daily, ginger appears to produce very few side effects.

To summarize, it is interesting to note that, after the initial studies of ginger's favorable effects on motion sickness, several additional investigations have been made to date. Half of these studies have given negative results and the other half have had positive outcomes. Factors that might explain this discrepancy include the failure to use standardized ginger preparations in the clinical trials. Ginger preparations vary widely in chemical composition and may contain inactive adulterants. Ginger grown in various areas of the world may also differ in content. It has also been well established that although there is no question about the physical basis of motion sickness, psychological factors also play an important role in suppressing as well as enhancing the tendency to be sick.

These problems should be resolved in future clinical trials by employing ginger preparations standardized in the amount of shogaols and gingerols now that they are known to be the antiemetic principles and by employing double-blind, crossover studies in the experimental design. In spite of these conflicting reports, analysis of some of the findings previously mentioned, together with other unpublished data, has been sufficient to convince Commission E of the German Federal Health Agency that ginger is effective not only for indigestion but also in preventing the symptoms of motion sickness. The commission recommends it at an average daily dose level of 2–4 g.[14]

Whatever activity ginger possesses may be attributed to a contained volatile oil (1–3 percent) responsible for its characteristic odor and oleoresin (mixture of volatile oil and resin) responsible for its pungency. The principal components of the volatile oil are the sesquiterpene hydrocarbons zingiberene and disabolene, which are accompanied by a number of other hydrocarbons and alcohols.[15] Nonvolatile pungent components include the shogaols and the gingerols, which have been shown to be the antiemetic principles by administration to leopard and ranid frogs, where emesis was induced by the administration of copper sulfate.[16]

Numerous references in the medical and the popular press suggest that ginger is useful in preventing or treating a variety of human ailments, including migraine headache, elevated cholesterol levels, rheumatism, hepatotoxicity, burns, peptic ulcers, depression, aging penile vascular changes, and impotence. Evidence supporting such claims is, as yet, insubstantial. The herb does have a long-standing and apparently valid reputation as a digestive aid.

Ginger has also been reported to be useful in the treatment of nausea of all kinds—not just that associated with motion sickness. One study has shown ginger to be as effective as metoclopramide in reducing the incidence of nausea and vomiting in a group of 120 women following elective laparoscopic gynecological surgery on a day-stay basis. The requirement for postoperative antiemetics was lower in patients who received ginger.[13] However, in a study involving 108 Australian patients undergoing elective gynecological laparoscopic surgery, it was reported that ginger in doses of 0.5 and 1.0 g was ineffective in reducing the incidence of postoperative nausea and vomiting.[14]

On the other hand, ginger's antiemetic action has been shown to be effective in treating hyperemesis gravidum, a severe form of pregnancy-related nausea and vomiting.[20] Vutyavanich and others[21] conducted a randomized controlled trial (RCT) with parallel design to examine the effectiveness of ginger powder against nausea and vomiting in sixty-seven pregnant women who were new obstetric attendees of a prenatal clinic. The patients after each of three daily meals for four consecutive days. A significant improvement was noted in the nausea scores of the ginger group compared to the placebo group, but it was only evident on the fourth day of treatment.

Ginger is ordinarily taken in the form of hard gelatin capsules, each containing 500 mg of the powdered rhizome. To prevent motion sickness, swallow two capsules thirty minutes before departure and then one or two more as symptoms begin to occur, probably about every four hours. A pleasant way to consume the product is in the form of candied or crystallized ginger, which is readily available in Asian food markets. This is prepared by boiling the fresh rhizome in a syrup solution, and the resulting product is marketed in slices sprinkled with granulated sugar. A piece 1 in. square and 0.25 in. thick weighs about 4 g. Because of the loss of some active principles during preparation and the presence of more water in this type of ginger, such a piece is probably equivalent to about one 500-mg capsule.

Backon has noted that ginger inhibits thromboxane synthetase and acts as a prostacyclin agonist. Thus, it could prolong bleeding time and produce immunological changes. In view of this, he suggests caution in using ginger to treat postoperative nausea.[22] However, although ginger inhibits platelet aggregation in vitro—thus acting as a potent inhibitor of arachadonic acid, epinephrine, adenosine diphosphate (ADP), and collagen—no bleeding episodes have been associated with its consumption. One case report associated consumption of a marmalade containing 15 percent raw ginger with inhibited platelet aggregation. Three clinical trials designed to look at bleeding risk have been performed with ginger, more than with any other herb. In one of these, a placebo-controlled trial, it was found that a single 10-g rather than a 4-g dose of powdered ginger significantly reduced platelet aggregation in patients with coronary artery disease.[23]

In addition to receiving approval of the German Commission E as previously noted, the herb appears on the GRAS (generally recognized as safe) list of the U.S. Food and Drug Administration.

Appetite loss

It has long been believed that the consumption of bitter herbs stimulates the appetite. When a bitter substance interacts with the taste buds at the base of the tongue, stimuli pass, primarily by way of the glossopharyngeal nerve, to a group of special cells in the cerebral cortex. The taste is interpreted there as bitter, and this causes stimuli to be forwarded through the vagus nerve to both the salivary glands and the stomach. The flows of saliva and of gastric juice are increased, as is the motility of the stomach. This stimulation of the digestive process enhances the appetite. Additional augmentation occurs when the bitter-tasting material actually reaches the stomach and promotes the secretion of gastrin, a hormone that intensifies the secretion of hydrochloric acid by the gastric glands.[24]

Two theories exist regarding the effectiveness of bitter substances as appetite stimulants; there is scanty experimental evidence supporting either.[25] The first holds that the stimulation of digestion by the administration of bitters to normal persons is greater than that induced by regular foods. The second postulates that such increases occur only in persons suffering from secretory deficiencies and that, in normal persons, no increased digestive capability is induced. Additional studies are required to determine the precise degree of utility of bitter herbs as appetite stimulants and digestion promoters in normal persons.

Nevertheless, the use of "bitter tonics" is widespread in Europe, where they are often consumed in the form of strong (about 40 percent) alcoholic extracts. Because alcohol is also a known appetite stimulant based on both its local irritating and CNS effects, it becomes difficult to separate its action from that of the contained bitter herbs. The use of bitter drugs to stimulate appetite is not rational in cases of chronic appetite loss that are manifestations of more severe conditions such as anorexia nervosa. It may be useful, however, in treating loss of appetite in elderly persons suffering from reduced production of stomach acid and digestive enzymes.

Significant bitter herbs

Gentian

One of the most popular of the bitter tonic herbs is gentian. It represents the dried rhizome and roots (underground parts) of *Gentiana lutea* L. (Gentianaceae), a moderately tall perennial herb with clusters of characteristic orange-yellow flowers. The best-quality product is dried quickly and retains its white color. Slowly dried gentian becomes reddish in color and develops a distinctive aroma but contains less of the desired bitter principles. However, even the well-prepared herb darkens and develops the distinctive odor after a period of six to eight months.[26]

The useful constituents of gentian are secoiridoid glycosides, principally gentiopicroside (gentiopicrin), swertiamarin, and especially amarogentin, which are primarily responsible for its bitter taste. Amarogentin, which occurs in gentian in a concentration of 0.05 percent, surpasses the bitter value of the more abundant gentiopicroside by a factor of five thousand. These principles are accompanied by a number of xanthones, alkaloids, phenolic acids, and characteristic sugars, such as the bitter gentiobiose.

The bitter taste of gentian is probably best known to most Americans in the form of angostura bitters, a proprietary cocktail flavoring that contains gentian, not angostura. Other vestiges in the United States of the numerous varieties of highly alcoholic stomach bitters that were once sold to stimulate the appetite and facilitate digestion are bitter aperitifs such as Campari and vermouth. However, the former relies on quinine for its bitterness and the latter on a complex mixture of herbal ingredients that varies with the producer but always includes wormwood. Wormwood is GRAS listed in the United States and may be used there as a flavor only if it is free of thujone, a toxic bicyclic terpene. In Europe, gentian aperitifs and liqueurs are both numerous and popular.

The traditional use of gentian as an appetite stimulant in some malnourished individuals, especially the elderly, is probably valid. German Commission E has reported that the bitter principles in gentian stimulate the taste buds and increase by reflex action the flow of saliva and stomach secretions. For this reason, it is said to act as a tonic.[27] Although animal studies have shown that gentian and its constituents may be potentially useful in the treatment of stomach ulcers,[28] that, as well as numerous other reported uses, requires verification.

The herb is probably best consumed in the form of a decoction prepared by gently boiling ½ level teaspoonful (ca. 1 g) of coarsely powdered root in ½ cup of water for five minutes. Strain and drink while still warm about thirty minutes before mealtime. If the beverage thus prepared is so strong as to be unpleasant, reduce the amount of herb accordingly. Consumption of gentian may be repeated up to a total of four times daily. The herb may cause headache in certain predisposed individuals.

Centaury

Another member of the Gentianaceae very similar to gentian in its constituents and effects is centaury. This herb consists of the dried aboveground parts—leaves, stems, and flowers—of *Centaurium erythraea* Rafn., a plant native to Europe and Asia but naturalized in the United States. Like gentian, it contains amarogentin, gentiopicroside, swertiamarin, and a number of related bitter principles including centapicrin, with a bitterness value fifteen times greater than gentiopicroside. It is used for precisely the same purpose as gentian, but Pahlow recommends that a heaping teaspoonful of the herb be extracted with a cup of *cold* water for six to ten hours, stirring occasionally, to prepare the most active beverage.[29] This "tea" should be warmed before drinking. The herb is approved by German Commission E for treatment of appetite loss and indigestion.[30]

Minor bitter herbs

Other, less important bitter herbs that may be encountered from time to time are discussed next.

Bitterstick

This consists of the dried plant *Swertia chirata* Buch.-Hamilt., family Gentianaceae.

Blessed thistle

This is the dried aboveground plant *Cnicus benedictus* L. (*Carbenia benedicta* Adans.), family Asteraceae.

Bogbean

The herb comprises the dried leaves of *Menyanthes trifoliata* L., family Menyanthaceae. None of the minor bitter herbs has any particular advantage over gentian or centaury, and they will not be discussed further here. Another bitter herb that cannot be recommended because of its potential toxicity is wormwood.

Wormwood

The dried leaves and flowering tops of *Artemisia absinthium* L. (family Asteraceae) were mentioned briefly as a constituent in vermouth. The herb will not be considered in detail because the volatile oil that it yields to the extent of 0.2–0.8 percent contains variable amounts (35 percent is not uncommon) of a toxic mixture of (–)-thujone and (+)-isothujone. Even though Wichtl has maintained that relatively little thujone is found in aqueous wormwood preparations (teas),[31] it does not seem prudent to employ a potentially poisonous herb as a bitter when other, safer ones are available.

Constipation

Clinicians generally define constipation as a decrease in the frequency of fecal elimination or difficulty in passing stools. In a survey of public perceptions of digestive health, 62 percent of American respondents believed that a bowel movement each day is necessary for good health. Whereas this frequency of bowel movement may be average, normal bowel habits may range between three and twenty-one stools per week. Fears of autointoxication by retention of noxious substances in the colon are unfounded if hepatic function is normal.

The decision to self-medicate for constipation depends largely on an individual's perception of abnormal bowel habits; however, in otherwise healthy individuals, laxatives should be considered of secondary importance to a fiber-rich diet, adequate fluid intake, and regular physical activity. In addition, laxatives are contraindicated in patients with cramps, colic, nausea, vomiting, bloating, and any undiagnosed abdominal pain.

Bulk-producing laxatives

When consumed with sufficient liquid, bulk-producing laxatives bind water in the colon, thereby softening the feces and expanding their bulk, which stimulates peristalsis and emptying of the bowel. They are probably the safest laxatives, functioning as they do in a fashion identical to high-residue foods rich in the dietary fiber found in plant cell walls. Dietary fiber is resistant to digestion and consists of varying quantities of cellulose, hemicellulose, pectin, lignin, cutin, waxes, and glycoproteins. Foods differ widely in the type and content of fiber they contain. Grains, cereals, and bran (more than 40 percent dietary fiber) contain large amounts of insoluble, poorly digested fiber, and their ingestion will shorten intestinal transit time and increase stool bulk. Fruits and vegetables contain more water-soluble fiber that results in a moister stool with less effect on transit time.[32]

Psyllium seed

Probably the most popular of the bulk-producing laxatives is psyllium seed, also known as plantago seed. It is estimated that psyllium-containing laxatives are used by four million Americans daily. Plantago seed is official in the *United States Pharmacopoeia XXII,* where it is described as the cleaned, dried, ripe seed of *Plantago psyllium* L. or of *Plantago indica* L., known in commerce as Spanish or French psyllium seed; or of *Plantago ovata* Forskal, known in commerce as blond psyllium or Indian plantago seed. All of these plants are members of the family Plantaginaceae.[33]

The seed coats or husks of plantago seeds contain cells filled with mucilage that is neither absorbed nor digested in the intestinal tract. In contact with water, it swells to a large volume, thus providing both bulk and lubrication and causing either the whole seed or the husks to act as an effective bulk-producing laxative. Recent clinical studies have demonstrated the effectiveness of plantago seed therapy. In one study of twenty-two subjects with idiopathic constipation, it was found that stool frequency increased significantly after eight weeks of treatment, as did stool weight. Subjects also reported an improvement in stool consistency and less pain on defecation.[34] In another study in older men and women with chronic constipation involving a single-blind, randomized, placebo-controlled fiber intervention with crossover, investigators found that psyllium fiber supplementation decreased total gut transit time from fifty-four hours to thirty hours. However, it did not rectify pelvic flow dysynergia, which was the principal cause of constipation in the group of patients studied.[35]

The usual dose of the seed is 7.5 g (2 heaping teaspoonfuls); 1 teaspoonful of the husks will provide the same effect. These doses are usually effective in twelve to twenty-four hours but may require as long as seventy-two hours in some individuals. Plantago seed husks, usually referred to as psyllium in the health-food industry, are readily available prepackaged and in bulk. For best results, stir the husks into a glass of water, juice, or milk and drink it quickly before the mixture has a chance to thicken.

In recent years, studies have shown that psyllium alone or mixed with other water-soluble dietary fibers in the diet of adult humans and children produces a modest but significant fall in total serum cholesterol and in low-density lipoprotein cholesterol.[36,37] A suggested mechanism is that psyllium increases fecal excretion of bile acids by binding bile acids in the intestine. Less is reabsorbed and less reenters the enterohepatic circulation; therefore, in the liver more cholesterol is converted into bile acids to replace bile acids lost in fecal excretion. The net result is to decrease the amount of cholesterol available for incorporation into circulatory lipoproteins.[38]

An abundance of evidence suggests that dietary fiber may reduce the risk of colon cancer. Dietary fiber can bind to carcinogens and tumor promoters, inhibiting carcinogenesis, and a preponderance of human epidemiological studies has demonstrated an inverse association between high-fiber diets and colon cancer. A recent study has convincingly demonstrated the role of wheat bran and psyllium in the chemoprevention of colon carcinogenesis in the rat experimental model.[39]

Stimulant laxatives

A number of herbal medicines producing a laxative effect contain mixtures of anthranoid compounds including glycosides of anthraquinones, anthranols, dianthrones, and anthrones. These herbs are classified as stimulant laxatives because they stimulate peristalsis via mucosal irritation or intraneural nerve plexus activity, which results in increased motility. However, what is even more important is their action on colonic mucosal cells resulting in the opening of Cl^- channels; this stimulates active Cl^- secretion with a net reduction of liquid and electrolyte absorption in the colon. The subsequent increase in water and electrolytes in the colonic lumen results in greater pressure in the intestine that leads to a laxative action.[40]

The anthranoid glycosides are essentially prodrugs. Upon oral administration, they pass to the colon unmodified, where they are hydrolyzed by the enzymes of the microbial flora of the gut to the free anthranoids plus glucose, rhamnose, or apiose. In addition, the free anthraquinones are inactive and are reduced by the microbial flora to the more active anthrones. Because the administered compounds must reach the colon to be activated, their effects are delayed and occur between six and twelve hours after ingestion. Therefore, a single bedtime dose promotes a morning bowel movement. These agents are indicated for constipation in patients who do not respond to milder drugs or have disorders such as anal fissures or hemorrhoids and after anal–rectal surgery in which an easy evacuation of the bowel with a soft stool is desired. They are also used for bowel evacuation before investigative procedures or surgery.[41]

Adverse effects of these medications include abdominal cramps (also known as griping), nausea, electrolyte disturbances (e.g., hypokalemia, hypocalcemia, metabolic acidosis, or alkalosis), and increased mucus secretion. Because of the loss of potassium (hypokalemia), the effect of digitalis cardiac glycosides may be potentiated. Another drawback to their use includes their tendency to promote overemptying and reduction of spontaneous bowel function, thus leading to development of the so-called laxative habit. For this reason, chronic use of stimulant laxatives should be discouraged and use beyond one week should be avoided.

Being absorbed into the general circulation, anthranoids find their way not only into the bile, urine, and saliva, but also into the milk of lactating women. Renal excretion of the compounds may cause abnormal coloration of the urine (yellowish brown that turns red with increasing pH). Large doses may produce nephritis, and when they are taken over extended periods, melanotic pigmentation of the colonic mucosa (melanosis coli) has been observed. This is reversible and thought to be benign, but its presence may help confirm a suspicion of laxative abuse.

In recent years, evidence has been reported that several anthraquinones are mutagenic in the Ames test. Some have been found to induce genotoxic effects in cultured mammalian cells and to behave as possible tumor promoters. As a consequence, in 1996 the FDA reclassified the OTC laxatives aloe, cascara sagrada, and senna from category I (safe and effective) to category III (more data needed); however, these drugs were allowed to stay on the market. Human clinical and epidemiological studies do not clearly suggest that use of anthraquinone herbal laxative drugs represents a risk for colorectal cancer or mutagenic effects under normal use conditions.[42,43] With this in mind, however, it would be prudent not to use these agents continuously over long periods of time.

Significant stimulant laxatives

Cascara sagrada

Of all the herbs that function as stimulants, the best is certainly cascara sagrada. This herb enjoys official status in the USP XXII, where it is described as the dried bark of *Rhamnus purshianus* DC. (family Rhamnaceae). Obtained from a small tree native to the Pacific Northwest, the bark should be collected at least one year prior to use in order to allow some of the harsh laxative, reduced emodin glycosides (anthrones) originally present in the bark to be oxidized naturally to less active monomeric anthraquinone glycosides. Cascara owes its action to a mixture of principles consisting largely of cascarosides A, B, C, and D, with other anthraquinone glycosides in minor amounts. Bark of USP quality contains not less than 7 percent total hydroxyanthracene derivatives calculated as cascaroside A on a dried basis. The cascarosides should make up at least 60 percent of this total.

Cascara is probably the mildest of the anthraquinone stimulant laxatives, producing only minor effects on the small intestine. Because of its relatively mild action, the herb is the least likely of the stimulant laxatives to produce undesirable side effects such as griping or dependence. Nevertheless, the active principles are excreted in mother's milk, so nursing mothers and, for that matter, pregnant women should avoid taking it or other anthraquinone-containing herbs.

Although cascara is normally taken in the form of prepared pharmaceutical dosage forms such as an extract, fluidextract, or aromatic
fluidextract, it is also possible to consume the powdered bark in capsule
form. Average dose is 1 g (about ½ teaspoonful). Cascara tea is not popular
because of its extremely bitter taste. The herb is an ingredient in several
popular OTC laxatives.[44]

Buckthorn (frangula) bark

This product is very similar to cascara because it is obtained from its
near-relative *Rhamnus frangula* L. (family Rhamnaceae), a shrub or small
tree that grows in Europe and western Asia. Its laxative effect is due to
the presence of anthraquinone derivatives, particularly glucofrangulin
A and B and frangulin A and B. Like cascara, buckthorn bark should
be aged one year prior to use in order to allow the reduced glycosides
(anthrones) with their harsh laxative action to be converted to milder
oxidized forms.[45]

Properly aged buckthorn bark is also comparable to cascara in its relatively gentle laxative action. In spite of this, it is not commonly used in
the United States, where it is overshadowed by the more popular native
species *Rhamnus purshianus*. It is a very popular drug in its native Europe.
Cascara is somewhat less expensive than buckthorn bark in the United
States, so there appears to be no particular advantage in using the latter.

The German Commission E has found buckthorn bark to be an effective stimulating laxative.[46] Average dose is 1 g. A fluidextract of the botanical, once official in the NF, remains a useful dosage form.

Senna

Probably the most widely used of the anthraquinone-containing stimulant laxatives is senna. Official in the USP XXII, senna is described there as
consisting of the dried leaflet of *Cassia acutifolia* Delile, known in commerce
as Alexandria senna, or of *Cassia angustifolia* Vahl, known in commerce as
Tinnevelly senna. These low-growing shrubs of the family Fabaceae are
native to Egypt or to the Middle East and India, respectively. Dianthrone
glycosides, particularly sennosides A, A_1, B, C, D, and G, together with
various other anthraquinone derivates, account for the laxative action of
senna.[47] The total complex of senna glycosides is also official in the USP
under the title "sennosides."

Although senna is not as mild in its action as cascara, producing
more smooth muscle contractions with attendant cramping, it is nevertheless more widely used because it is considerably cheaper. Bulk lots
of the herb are only about one-half the price of cascara. A fluidextract
and a syrup made from the leaflets are available, as are tablets prepared
from a mixture of the purified active ingredients—so-called sennosides.
A bitter-tasting tea can be prepared from 0.5–2 g (0.5–1 teaspoonful) of

the herb. Some prefer a beverage prepared by soaking the leaflets in cold water for ten to twelve hours and then straining. Such a preparation will be more active than the customary hot tea and will contain less resinous material.[48]

Other stimulant laxatives

Other laxative herbs in this category that are encountered with some frequency are discussed next.

Aloe (aloes)

Also official in the USP XXII, the drug consists of the dried latex of several species of aloe, especially *A. barbadensis* Mill. (*A. vera* L.), known in commerce as Curaçao aloe, or of *A. ferox* Mill., and hybrids of these species with *A. africana* Mill. and *A. spicata* Bak., known in commerce as Cape aloe. All of these species are members of the family Liliaceae. The anthraquinone glycosides aloin A and B (formerly designated barbaloin) render aloe an extremely potent laxative. Although still widely used abroad, its use in the United States has declined in recent years, rendering it of relatively minor therapeutic significance.

Aloe is obtained from specialized cells, known as pericyclic tubules, that occur at the border of the outer and inner cortical layers of the mesophyll, located just beneath the epidermis of the leaves of the aloe plant. The bitter yellow latex found there is drained and dried to a reddish black glistening mass. This drug is totally different from the colorless mucilaginous gel (aloe gel or aloe vera gel) obtained from the parenchyma tissue making up the central portion of the aloe leaf. Aloe gel is primarily a wound-healing agent and is discussed in Chapter 10; it should never be confused with aloe, the laxative.

Rhubarb

This consists of the dried rhizome and root of *Rheum officinale* Baill., *R. palmatum* L., related species, hybrids grown in China (Chinese rhubarb) and Japan (*R. palmatum* x *R. coreanum* Nakai, Japanese rhubarb), or *R. emodi* Wall or *R. webbianum* Royle, native to India, Pakistan, or Nepal (Indian rhubarb). It is not common garden rhubarb, *R.* x *cultorum* Hort. (*R. rhaponticum* Willd.), which is relatively inactive.

As is the case with aloe, this laxative herb is much more potent than cascara or senna. Its use almost always causes intestinal griping or colic. For this reason, rhubarb is seldom employed as a laxative today and cannot be considered an herb of choice. A very popular "tonic" preparation, widely sold in Europe and occasionally marketed in the United States, is Swedish bitters. Its active constituents include senna, aloe, and rhubarb; frequent use is not recommended.[49]

There are many other laxative herbs. Some, such as colocynth, jalap, and podophyllum, are drastic purgatives; others, such as dandelion root and manna, are so mild as to be uncertain in their action. None provides any advantage over psyllium seed, cascara, buckthorn bark, or senna.

Diarrhea

Diarrhea is characterized by increased fluidity and volume of the stool due to the excretion of a relatively large volume of water normally reabsorbed from the gut and almost always accompanied by increased frequency of bowel movement. Although normal functioning of the bowels varies greatly in different individuals, which must be taken into account in the definition, diarrhea ordinarily involves defecation more than three times daily. The condition may be acute or chronic, and the causes are numerous, ranging from allergies to viral infections.

Acute diarrhea is characterized by a sudden onset of self-limited episodes of less than forty-eight hours' duration; if it persists for more than three to four days, a physician should be consulted. It can usually be attributed to intestinal infection caused by various organisms such as *Salmonella*, invasive *Escherichia coli*, or enterovirus. Some organisms, such as *Staphylococcus aureus* and toxigenic strains of *E. coli*, cause diarrhea through food contaminated with an enterotoxin (food poisoning). Food-induced diarrhea may also be caused by food allergy or by ingestion of foods that are excessively spicy or high in roughage. In addition, certain drugs may induce diarrhea as an adverse effect.

Chronic diarrhea is characterized by loose stools over a prolonged period, and it may be a symptom of a serious disorder of the gastrointestinal tract. Individuals who experience persistent or recurrent diarrhea for more than two weeks and are unaware of the cause should seek prompt medical attention from a physician.

Several preparations available as OTC drugs are effective in the treatment of diarrhea. These include loperamide (Imodium), bismuth subsalicylate (Pepto-Bismol), and kaolin pectin suspension (Kaopectate). Herbal remedies are also available and are effective due to their content of polyphenolic substances, popularly known as tannins.

These compounds tend to arrest diarrhea by their astringent action, which reduces intestinal inflammation. They effect this by binding to the surface protein layer of the inflamed mucous membranes, causing it to thicken and thereby hindering resorption of toxic materials and restricting secretions. Tannins that are slowly released from the various complexes in which they exist in the plant after reaching the lower gastrointestinal tract are most effective.[50] This prevents stomach upset and also allows the tannin to act at the site where it is most effective. The most widely used astringent herbs are obtained from several edible berry plants.

Because of their similarities, the leaves of three plants may be considered as a group:

- blackberry leaves—the dried leaf of *Rubus fruticosus* L. (family Rosaceae);
- blueberry leaves—the dried leaf of *Vaccinium corymbosum* L. or *V. myrtillus* L. (family Ericaceae); and
- raspberry leaves—the dried leaf of *Rubus idaeus* L. or of *R. strigosus* Michx. (family Rosaceae).

All of the leaves contain appreciable amounts of tannin; quantities up to 6.7 percent have been reported for blueberry leaves[51] and a range of 8 to possibly 14 percent has been reported for blackberry leaves.[52] All three are consumed in the form of teas, usually prepared by pouring a cup of boiling water over 1–2 teaspoons of the finely cut leaves and steeping for ten to fifteen minutes. Alternatively, the plant material may be macerated in cold water for about two hours and then strained to yield the beverage. A cup of the tea may be drunk up to six times a day as necessary to control the diarrhea. However, if the condition lasts more than two or three days, it is obviously not amenable to treatment with astringent herbs. The teas from all three berry leaves may also be used effectively as a mouthwash or gargle for sore mouth and inflammation of the mucous membranes of the throat.

Raspberry leaf tea has a persistent traditional reputation of utility in treating a wide variety of female conditions ranging from diabetes and menstrual difficulties to those associated with pregnancy, such as morning sickness and labor pains. One reputable scientific reference continues to note that the tea has been a traditional remedy for "painful and profuse menstruation and for use before and during confinement."[53] The German Commission E concluded that the leaf has not been proven effective for any of these complaints[54] (see Chapter 8 for additional details). Other useful antidiarrheal herbs are discussed next.

Other antidiarrheal herbs

Blackberry root

Also known by the Latin title rubus, this was listed in the *National Formulary* until 1936, where it was defined as the dried bark from the rhizome and roots of plants of the section *Eubatus* Focke of the genus *Rubus* L. (family Rosaceae). The section *Eubatus* includes more than fifty species and cultivated varieties of the plants commonly referred to as blackberries and dewberries. It is of little value to enumerate them here because they are all similar qualitatively and quantitatively in their tannin content and are used similarly.

Although it appears on the GRAS list, the root bark no longer seems to be a common article of commerce in the United States. It may be self-collected and dried, in which case it is best used by extracting the finely powdered or cut bark in boiling water for twenty minutes because the tannins are more difficult to extract from bark than from leaf material. Blackberry root bark is an effective antidiarrheal agent, but the traditional use of the root as a preventative treatment for dropsy is unproven; it cannot be recommended for this latter purpose.[55]

Blueberries

Dried blueberry fruits (*Vaccinium corymbosum* L. in the United States and *V. myrtillus* L. in Europe) are highly recommended, particularly by European authorities, to combat simple diarrhea. They do not appear to be an item of commerce in the United States but may be prepared by drying fresh berries in the sun. Do not use the fresh blueberries themselves for treatment of diarrhea. They may even exert a laxative effect.[51]

Use of the dried berries involves chewing three to five tablespoonfuls and swallowing them. Alternatively, a drink may be prepared by boiling the crushed fruits in water for about ten minutes, straining, and drinking. In addition to tannins, the berries contain pectin, which, by virtue of its adsorbent properties, is also useful in the treatment of diarrhea.

There are several other reputed uses of these fruits, particularly of *V. myrtillus*, which are often referred to in the American herbal literature as bilberries. Preliminary studies indicate that a concentrated extract of anthocyanosides obtained from these berries may benefit visual acuity as well as provide protection against macular degeneration, glaucoma, cataracts, and the like.[56] It must be emphasized that these claims are based on studies carried out primarily in small animals or on small numbers of human subjects. Extensive clinical studies in human beings are required to verify these initial findings before the herb can be recommended for these purposes. Dried bilberries nevertheless remain a useful treatment for diarrhea. In the meantime, a bilberry extract containing 25 percent anthocyanosides calculated as anthocyanidin is widely sold as a phytomedicine in Europe, and it is available in the United States as a food supplement.

Indigestion—dyspepsia

In modern medicine, digestive disturbances are treated with a variety of drugs, ranging from antacids and H_2-receptor antagonists in the case of acid peptic disorders to synthetic antispasmodics for functional bowel complaints and cramps. In herbal medicine, stomach upsets, bloating, and related digestive problems are customarily treated with certain plant drugs, usually those containing volatile oils, that act as carminatives. Narrowly defined, a carminative is an agent that relieves flatus (gas in the stomach or intestine). But present-day usage attributes a much broader range of action to such drugs.

Carminatives

According to Schilcher, carminative effects on the stomach, gall bladder, and intestinal tract result from at least five different activities; the intensity of each varies, depending on the identity of the specific herb.[57] These effects include:

- local stimulation of the stomach lining, leading initially to an increase in tonus and an intensification of rhythmic contractions facilitating the eructation (belching) of air from that organ; this is promoted by relaxation of the lower esophageal sphincter;
- reflexive increase in stomach secretions, resulting in improved digestion;
- antispasmodic or spasmolytic effects on smooth muscle; the intestine is especially relaxed, facilitating the passage of intestinal gas;
- antiseptic action, limiting the development of undesirable microorganisms; and
- promotion of bile flow (cholagogue effect), facilitating digestion and absorption of nutrients.

Determination of the effectiveness of herbs containing volatile oil in treating gastrointestinal problems is difficult, and the results are controversial. In a series of in vitro experiments, Forster and colleagues concluded that the antispasmodic activities of the most effective plant extracts tested—including peppermint, chamomile, and caraway—were less than those produced by atropine.[58] Hof-Mussler has noted that observations of the effects of carminative herbs on isolated smooth muscle preparations have not necessarily been confirmed by subsequent in vivo experiments.[59]

Clinical studies of the spasmolytic action of volatile oils have yielded mixed results. To be effective, relatively high concentrations are required at the site of action. There is little doubt that, in many cases, the observed activity of these herbs in alleviating indigestion is due in part, but not entirely, to the placebo effect. There is also little question that they are effective, at least subjectively. Persons with gastric distress do feel better after consuming them. For that reason, they continue to be used to reduce such discomfort.

The mode of consumption is of considerable importance. In the United States, teas are the most common dosage forms. But volatile oils are relatively insoluble in water, so teas are not very efficient therapeutic agents. For example, it has been estimated that chamomile tea contains only about 10–15 percent of the volatile oil present in the plant material.[60] Fluidextracts or tinctures of the herbs prepared with 30–70 percent alcohol are much more effective, but they are not always available commercially and must be self-prepared. Alcoholic solutions of the purified volatile oils are also employed for the same purpose.

Significant carminative herbs

Peppermint

Although many plant materials contain volatile oils and therefore possess some carminative properties, one of the most effective and widely used is certainly peppermint. Consisting of the leaves and flowering tops of *Mentha* x *piperita* L. (family Lamiaceae), this herb is officially listed in the NF XVII. It contains 0.5–4 percent (average of about 1.5 percent) of a volatile oil composed of 50–78 percent free (*)-menthol and from 5 to 20 percent menthol combined in various esters such as the acetate or isovalerate. It also contains (+)- and (*)-menthone, (+)-isomenthone, (+)-neomenthone, (+)-menthofuran, and eucalyptol, as well as other monoterpenes.[61] Although flavonoid pigments found in the leaf may also exert some physiological effects,[62] there is little question that most of the activity is due to the constituents of the oil, primarily menthol.

Peppermint oil has long been an extremely popular flavoring agent in products ranging from chewing gum to after-dinner mints. It is probably the most widely used carminative, acting in the broad sense defined by Schilcher.[57] The German Commission E has found peppermint or its volatile oil to be effective as a spasmolytic (particularly for discomfort caused by spasms in the upper digestive tract), a stimulant of the flow of bile, an antibacterial, and a promoter of gastric secretions.[63] On the other hand, in 1990, the U.S. Food and Drug Administration declared peppermint oil to be ineffective as a digestive aid and banned its use as a nonprescription drug for this purpose.[64] What this actually means is that the FDA was not presented with evidence proving the efficacy of peppermint as a digestive aid. As previously explained, this would not be financially feasible in the United States. It does not mean that peppermint oil is an ineffective aid to digestion.

One of the most frequent diagnoses made for patients because of gastrointestinal complaints is non-ulcer dyspepsia. It is characterized by motility disturbances with bloated abdomen, a feeling of fullness, diffuse pain, nausea, vomiting, and intolerance of foodstuffs. The efficacy and safety of the herbal combination of peppermint oil (90 mg) and caraway oil (50 mg) in an enteric-coated capsule have been studied in a double-blind, placebo-controlled multicenter trial in patients with non-ulcer dyspepsia. After four weeks of treatment, the group of patients treated with the herbal combination had improved significantly with few adverse effects.[65] Peppermint oil is also used to relieve the symptoms of irritable bowel syndrome, and experimental evidence indicates that it acts by relaxing intestinal smooth muscle by reducing calcium availability in the muscle membrane.[66]

Peppermint is GRAS listed, and both it and peppermint oil are recognized as flavoring agents in the NF XVII. Peppermint tea is prepared by pouring about 2/3 cup of boiling water over a tablespoonful of the recently dried leaves and steeping for five to ten minutes. Drink this amount of tea three to four times daily between meals to relieve upset stomach. Peppermint spirit (USP XXII), an alcoholic solution containing 10 percent peppermint oil and 1 percent peppermint leaf extract, is also available in pharmacies. The usual dose is 1 mL (20 drops) taken with water.

Regular consumption of peppermint tea is considered safe for normal persons, although excessive use of the volatile oil (0.3 g = 12 drops) may produce some toxic effects.[67] Allergic reactions to menthol have also been reported.[67–69] Peppermint tea should not be given to infants or very small children because they often experience an unpleasant choking sensation due to the menthol.[70]

Chamomile

Another extremely popular herb for the treatment of indigestion, as well as various other conditions, is chamomile. German or Hungarian chamomile, often referred to as true chamomile or matricaria, consists of the dried flower heads of a plant now technically designated *Matricaria recutita* L. In the older literature on medicinal plants, it is usually referred to as *Matricaria chamomilla* L.p.p. or as *Chamomilla recutita* (L.) Rauschert. At least seven other scientific names have been used to designate the plant. A member of the daisy family and formerly known as the Compositae, it is now called the Asteraceae by most taxonomists. The important thing to remember when attempting to maneuver through this nomenclatural maze of five English common names, ten scientific names, and two plant family designations is that all of the names refer to a single species of plant. Its popularity as a traditional medicine is so great that in Germany it was declared the medicinal plant of the year for 1987.[71]

A related plant is Roman or English chamomile, composed of the flower heads of *Chamaemelum nobile* (L.) All. syn. *Anthemis nobilis* L. (family Asteraceae). Its constituents are not identical to those of German chamomile, but both plants are similarly employed. In Great Britain, Roman chamomile is the herb of choice; on the Continent, German chamomile is preferred. The latter is also the species most commonly consumed in the United States.

The literature on the chamomiles is extensive. *Die Kamille,* a 152-page book by H. Schilcher, provides comprehensive coverage.[72] Mann and Staba's 1986 review is the most recent one in the English language.[73] It provides 220 references to various agronomic, botanical, chemical, and pharmacological aspects of the herbs.

A blue-colored volatile oil is obtained from the chamomiles by steam distillation in yields up to nearly 2 percent. The blue color is due to chamazulene, which is formed during distillation from matricin, a sesquiterpene found in chamomile flowers. Chamazulene was shown to have anti-inflammatory action as early as the 1930s. Its precursor, matricin, is also an active anti-inflammatory. In the 1950s, it was found that the constituent flavonoids, especially apigenin, have both anti-inflammatory and antispasmodic activities. Later it was shown that the terpenoids (–)-α-bisabolol and (–)-α-bisabololoxides A and B have similar activities.[74] Della Loggia and others, in 1990, determined that topically the most active anti-inflammatory principle in chamomile is apigenin, followed by matiricn, (–)-α-bisabolol, and chamazulene, with respective ID_{25} values of 12, 79, 105, and 1,080 $\mu g/cm^2$.[75] Chamazulene and bisabolol are very unstable and are best preserved in an alcoholic tincture; the essential oil exhibited the weakest inhibition of edema, and an alcoholic extract was far superior.

The chamomiles are used internally for digestive disturbances. Like peppermint, they possess carminative (antispasmodic) effects; however, unlike peppermint, they exert a pronounced anti-inflammatory activity on the gastrointestinal tract. In the United States, chamomile tea is most widely used for this purpose. It is prepared by pouring boiling water over a heaping tablespoonful (about 3 g) of the flower heads and allowing them to steep in a covered vessel for ten to fifteen minutes. A cup of freshly prepared tea is drunk between meals three or four times daily for stomach or intestinal disturbances. As previously noted, only a small fraction of the volatile oil (roughly 10 percent) and about 30 percent of the flavonoids are extracted in this way, but there is reason to believe the tea provides a cumulative beneficial effect.

Both German and Roman chamomile are considered safe for normal human consumption and are GRAS listed by the FDA. Some allergies have been reported, however, so persons with known sensitivities to them or to various other members of the Asteraceae (e.g., ragweed, asters, chrysanthemums) should be cautious about consuming chamomile teas. However, this rather remote possibility may have been greatly overemphasized in some of the nonmedical literature.[76] Only five cases of allergy specifically attributed to German chamomile were identified worldwide between 1887 and 1982.[77] However, a recent report indicates that a German chamomile ether extract used in allergic patch testing from 1985 to 1990 in 3,851 tested individuals produced an allergic reaction in sixty-six patients, or 1.7 percent.[78]

Most of the other forty-plus reports of allergic reaction probably relate to the common adulterant/substitute *Anthemis cotula* L. (stinking mayweed, dog's chamomile, dog fennel), which contains the particularly noxious sesquiterpene, anthecotulid.[79] In an alarming case of professional botanical misidentification, botanists at the University of Texas, Austin, vouchered a plant as *M. chamomilla* of Argentinean origin containing 7.3 percent anthecotulid, which was subsequently determined to be *A. cotula*.[77] The German Commission E considers German chamomile effective internally for gastrointestinal (GI) spasm and inflammatory conditions of the GI tract. It was also found to be effective when used as a mouthwash for irritations and minor infections of the mouth and gums.

In Europe today, a large number of pharmaceutical preparations are available containing either extracts of chamomile or chamomile volatile oil. Some are creams or lotions intended to treat various skin irritations, including those caused by bacterial infections. Other forms are suitable for inhalation and are designed to relieve bronchial irritation; still others are used as baths or rinses to alleviate irritations in the anogenital regions. Commission E has concluded that German chamomile is also effective for these purposes.[80] However, because none of these preparations has been approved for use as an OTC drug in the United States and such dosage forms are not ordinarily available there, their therapeutic use will not be discussed further.

Finally, it should be noted that chamomile herb is not inexpensive and is relatively easy to adulterate. To ensure quality, purchase it only in the form of the whole flower heads, which are easy to identify with a little experience, and make certain that no appreciable quantity of stems is present (less than 10 percent). It is best to avoid pulverized or powdered chamomile, the quality of which is difficult to determine, even by experts. Any preparations containing chamomile oil (most will be foreign made) should be acquired only from firms having outstanding reputations for quality products. Three-quarters of the commercial chamomile oil samples examined in 1987 were found to be adulterated, often with cheaper, synthetically prepared, blue-colored compounds such as guaiazulene.[81]

A natural source of chamazulene is the cheaper essential oil of yarrow (*Achillea millefolium* L.) and the wood of the Brazilian tree *Vanillosmopsis erythropappa* Schultz Bip., which yields up to 3 percent essential oil consisting almost exclusively of (–)-α-bisabolol, which has been used as a cheaper source of that terpenoid. Chamazulene from *A. millefolium* and bisabolol from *V. erythropappa* can only be distinguished from their corresponding chamomile constituents by isotope ratio mass spectrometry.[82]

Minor carminative herbs

Other herbs frequently used as digestive aids for the carminative properties of their contained volatile oils are discussed next.

Anise

This is the dried, ripe fruit of *Pimpinella anisum* L., family Apiaceae. The fruit is commonly referred to in the popular literature as a seed.

Caraway

This herb consists of the dried, ripe fruit (often called the seed) of *Carum carvi* L., family Apiaceae.

Coriander

This herb is composed of the dried, ripe fruit (often called the seed) of *Coriandrum sativum* L., family Apiaceae. Two varieties of this species are commonly employed: var. *vulgare* Alef. and var. *microcarpum* DC.

Fennel

This is the dried, ripe fruit (often called the seed) of various cultivated varieties of *Foeniculum vulgare* Mill., family Apiaceae. Three of these—anise, caraway, and fennel—are GRAS listed, and another group of three—anise, caraway, and coriander—was found to be effective treatment for indigestion by the German Commission E.[83] However, all have relatively weak activities and offer no particular advantage over either peppermint or chamomile as digestive aids except in the uncommon case of an allergy to one of these latter herbs. Each of the herbs is administered in the form of a tea prepared by pouring a cup of boiling water over 1 teaspoonful of the crushed fruit, steeping for ten to fifteen minutes, and straining prior to drinking.

An herb that has been used for centuries to relieve digestive disorders but that cannot be recommended on the basis of present-day knowledge is calamus.

Calamus

The aromatic rhizome (underground stem) of *Acorus calamus* L., family Araceae, constitutes this drug. At least four subtypes of this plant exist, two of which contain large amounts of β-asarone (*cis*-isoasarone), a compound found in experiments to promote the development of malignant tumors in the intestines of rats. A third type of calamus contains much less of this toxic compound, and a fourth (drug type 1), native to North America, is β-asarone free. However, these types are not normally differentiated in commerce; consequently, the use of calamus and its extracts is prohibited in the United States.[84]

Cholagogues

Pain of obscure origin in the upper portion of the stomach, frequently intensified by the consumption of fatty foods and accompanied by a feeling of fullness or bloating, is often attributed to an inadequate flow of bile from the liver. Traditionally, dyspepsia of this sort has been treated with herbal cholagogues—that is, agents acting to empty the gallbladder (cholekinetics), to stimulate the production of bile (choleretics), or both.

Hänsel has noted that many clinicians consider the use of such agents obsolete, and the terms themselves are not utilized in modern gastroenterology. Nevertheless, cholagogue phytomedicinals are widely used and do provide relief from this type of indigestion. They may also play some role in the prevention of gallstone formation, although evidence in support of this effect is less convincing.[85]

Herbs previously mentioned in this chapter that also function as useful cholagogues include gentian and peppermint. Although horehound possesses choleretic properties, its principal use is as an expectorant, so it is discussed in Chapter 5. As is the case with some other herbs in this category, the activity of these three plants in promoting bile flow is relatively mild, and they are considered to be both safe and effective.

Turmeric

Of the remaining herbal cholagogues, perhaps the best known and most widely used is turmeric. Consisting of the rhizome of *Curcuma longa* L. (synonym *C. domestica* Val.) of the family Zingiberaceae, turmeric has long been used as a yellow food coloring and spice. It is one of the principal ingredients in curry powder. Related species, such as *C. zanthorrhiza* Roxb. (Javanese yellow root) and *C. zedoaria* (Christm.) Rosc., commonly known as zedoary, have similar, but quantitatively different, therapeutic properties due to different concentrations of identical or closely related constituents.

In addition to 4.2–14.5 percent of an essential oil consisting of about 60 percent of sesquiterpene ketones (known as turmerones), turmeric contains three major curcuminoids, of which curcumin (diferuloylmethane) is the most significant, present at up to 4 percent in *C. longa*.[86] These curcuminoids are responsible for the yellow color of the herb.

The choleretic action of turmeric and related species is attributed primarily to its volatile oil; the cholekinetic effects, as well as appreciable anti-inflammatory properties, are believed to be due to the curcuminoids.[87] Animal studies have also demonstrated a hepatoprotective effect and antimutagenic and anticarcinogenic activity for turmeric,[88] but evidence to support its use in cases of liver disease or as a cancer preventive in humans is not yet available. Tumeric and curcumin have been found to reverse aflatoxin-induced liver damage in vivo and to inhibit aflatoxin production by *Aspergillus parasiticus*.[89] Supporting rodent studies demonstrated reduction of serum total and LDL cholesterol levels, as well as elevation of HDL cholesterol. In a study with human volunteers, curcumin was found to decrease the levels of serum lipid peroxides and total serum choleserol, as well as to increase HDL cholesterol.[90]

Attention is being increasingly directed to the antioxidant activity of turmeric. Turmeric is as effective an antioxidant as butylated hydroxyanisole (BHA), widely used in the preservation of foods. However, although high concentrations of BHA have been shown to be toxic, comparably high concentrations of tumeric are reported to be nontoxic. A heat-stable antioxidant principle, a protein termed TAP (tumeric antioxidant protein), has been isolated from an aqueous extract of *C. longa;* trypsin treatment abolishes the antioxidant activity.[91]

Curcumin is not well absorbed following oral administration; only traces subsequently appear in the blood when turmeric is given by mouth. However, the compound is active at relatively low concentrations, and it may also produce a local action in the gastrointestinal tract. The relatively poor absorption of the curcuminoids would seem to emphasize the role of the volatile oil constituents in the activity of turmeric.

Regardless of the identity of its active constituents, turmeric is a widely used and apparently effective cholagogue and digestive aid. Both it and Javanese yellow root have been declared effective for such use by the German Commission E.[92] Because both the contained volatile oil and the curcuminoids are relatively water insoluble, teas for therapeutic purposes are seldom made from turmeric.[93] Instead, hydroalcoholic fluidextracts or tinctures of the herb or encapsulated powders are the dosage forms customarily employed. The usual dose is 1.5–3.0 g of turmeric daily; preparations are consumed in equivalent quantities. Use over an extended period of time infrequently results in gastric disturbances. Persons suffering from gallstones or blockage of the bile duct should avoid consuming turmeric or even curry powder (28 percent turmeric).

Boldo

The dried leaves of *Peumus boldus* Mol. (family Monimiaceae), an evergreen shrub native to Chile, have an ancient reputation as a "hepatic tonic," diuretic, and laxative. The herb contains about 2 percent of a volatile oil, the principal constituents of which are ascaridole, eucalyptol, and *p*-cymol. About 0.25–0.5 percent of an alkaloidal mixture is also present. Boldine, an aporphine alkaloid, constitutes about one-fourth of the total; the remainder consists of about sixteen different alkaloids.[85]

Boldine is responsible for both the choleretic and diuretic activity of the leaves.[94] Although the herb may increase the flow of urine substantially, the mechanism of action is still unknown. Thus, it is uncertain if boldine's action is one of true diuresis or simply aquaresis, as is the case with most herbs in this category (see Chapter 4).

Although Commission E has approved the use of boldo for the treatment of dyspepsia as well as for stomach and intestinal cramps, it must be noted that the volatile oil in the leaves contains about 40 percent ascaridole, a rather toxic component. Because no chronic toxicity testing has been carried out, Hänsel has recommended that prolonged use of the herb or any consumption by pregnant women be avoided. Boldo is normally taken as a tea prepared from 1–2 teaspoonfuls of the herb. The average choleretic dose is 3 g daily.[95]

Dandelion

The dried rhizome and roots of *Taraxacum officinale* Weber (family Asteraceae) are mentioned here primarily as a matter of record. Otherwise, some readers might believe this common herb, usually known as dandelion root, has been overlooked. It has not been. Scientific evidence to support claims of any significant effects for this centuries-old home remedy is simply insufficient to allow it to be considered useful.

Dandelion root apparently possesses very mild choleretic properties, and the leaf is thought to be a feeble diuretic (aquaretic). However, most of the animal and clinical investigations on which these claims are based date from the 1930s.[96] The root, together with the aboveground parts of the plant, is said by the German Commission E to have choleretic, diuretic, and appetite-stimulating properties,[97] but additional verification is required.

Some authorities believe that the use of dandelion root as a choleretic stems principally from the fact that the plant has a yellow flower, so, according to the doctrine of signatures, it would be a useful remedy for jaundice. The plant lacks any significant documented pharmacological activity.[98]

Hepatotoxicity (liver damage)

Liver damage can result from certain disease states, including infections, vascular disorders, and metabolic disease, or by the consumption of hepatotoxins such as alcohol, certain drugs and chemicals, or deadly amanita mushroom poisons.

Chronic damage to the liver results in the condition termed cirrhosis. It is characterized by a diffuse increase in the fibrous connective tissue of the liver, with areas of both necrosis and regeneration of parenchymal cells imparting a nodular texture and hardening of the liver. In the United States, cirrhosis follows only cardiovascular disease and cancer as a cause of death in the forty-five to sixty-five age group, with most cases a result of chronic alcohol abuse. In many parts of Asia and Africa cirrhosis due to chronic viral hepatitis B is a major cause of death.

In general, the treatment of liver damage is supportive, with the exception that the recent availability of liver transplantation for patients with advanced cirrhosis or acute life-threatening liver damage from toxins has been life saving. Supportive treatment involves withdrawal of toxic agents, attention to nutrition (including supplemental vitamins), and the treatment of complications as they arise. Specific therapies focusing on the alteration of fibrous connective tissue and collagen formation are experimental and still being evaluated. Corticosteroids and penicillamine are two drugs in this category that have uncertain value.

Four herbal medicinal preparations commonly used in the treatment of liver disease—namely, *Phyllanthus amarus* Schum. (family Euphorbiaceae), *Silybum marianum* (L.) Gaertn. (family Asteraceae), *Glycyrrhiza glabra* L. (family Fabaceae), and Liv-52, a combination Ayurvedic herbal preparation—have been reviewed.[99] *Phyllanthus* showed a positive effect on clearance of hepatitis B viral (HBV) markers, and no major adverse effects were evident in consideration of twenty-two RCTs (*n* = 1,947) (thirteen monopreparations and nine combination products). However, due to the low methodological quality of most of these trials and limited follow-up, *Phyllanthus* was not recommended for clinical use in patients with chronic HBV infection until larger-scale prospective multicenter, randomized controlled trials demonstrate consistent benefit. Recently, a potent ex vivo anti-HIV activity was detected in sera of volunteers after administration of a *P. amarus* preparation.[100]

Although milk thistle extract appears safe and well tolerated, it does not reduce mortality among patients with chronic liver disease and improves neither biochemical markers nor histology at biopsy among such patients. As such, a recommendation of such preparations for treatment of liver disease is not warranted.

The licorice preparation here evaluated is actually an aqueous extract of *G. glabra* root containing 0.2 percent glycyrrhizin, combined with 0.1 percent cysteine and 2 percent glycine, a Japanese product termed "stronger neominophagen C" (SNMC). RCTs with SNMC indicate that it does not have antiviral properties, acting primarily as an anti-inflammatory or cytoprotective drug. It improves mortality in patients with subacute liver failure, compared to historical controls, but does not prevent chronic sequelae in uncontrolled experiments. However, it appears to improve liver function in patients with subacute hepatic failure and to prevent development of hepatocellular carcinoma in patients with chronic hepatitis C in similar experiments.

Liv-52 has been judged "not useful in the management of alcohol-induced liver disease." However, a contemporary RCT concluded that the combination herbal product possesses hepatoprotective effect in cirrhotic patients, attributable to diuretic, anti-inflammatory, antioxidative, and immunomodulating properties of the component herbs.[101]

Herbal liver protectants

Milk thistle

At least one herbal remedy has shown considerable promise as a liver protectant in conditions of this sort. Milk thistle has been widely used and extensively investigated as a cure or preventive for a wide range of liver problems. This herb consists of the ripe fruits, freed from their pappus (tuft of silky hairs), of *Silybum marianum* (L.) Gaertn. (family Asteraceae). A crude mixture of antihepatotoxic principles was first isolated from the plant and designated silymarin. It is contained in the fruit in concentrations ranging from 1 to 4 percent. Subsequently, silymarin was shown to consist of a large number of flavonolignans, including principally silybin accompanied by isosilybin, dehydrosilybin, silydianin, silychristin, and possibly several others, depending on the variety examined.[102]

Studies in small animals have shown conclusively that silymarin exerts a liver-protective effect against a variety of toxins, including those of the deadly amanita. Silybin has been investigated in several studies involving humans poisoned with the death cap mushroom, *Amanita phalloides*. It was shown to be an effective prophylactic measure against liver damage if it is administered intravenously within twenty-four hours after mushroom ingestion; however, the rule here is "the sooner the better," because there is a brief window of time after the ingestion of the poisonous cyclic peptides found in the mushroom before they are taken up by the liver. If the silybin can bind to receptor sites on the outer liver cell membranes targeted by these toxins before the toxins bind, it can prevent the entry of the toxins into the cell and reduce the severe liver damage that can result in death.[103]

Central to the use of milk thistle in the treatment of degenerative liver conditions is the ability of silymarin to increase protein synthesis in hepatocytes. These liver parenchymal cells make up the bulk of the organ and carry out the exquisitely complex metabolic processes responsible for the liver's essential role in metabolism. The biochemical mechanism by which this increased protein synthesis is achieved is illustrated in Figure 3.1.

Silymarin stimulates RNA polymerase I activity, possibly by activating a promoter or enhancer site on the DNA that the polymerase uses as a template for the synthesis of ribosomal RNA. Ribosomal RNA is required for the formation of ribosomes in the cytoplasm, where protein synthesis takes place. Therefore, an increase in the amount of ribosomal RNA results in an increased number of ribosomes, with a resulting increase in protein synthesis (enzyme and structural protein). The increase in enzymes and structural protein stimulates the repair of injured cells and also increases the synthesis of DNA, resulting in an increase in mitosis and cell proliferation.[104,105] In addition, silymarin functions as a free-radical scavenger and antioxidant.[106]

The German Commission E endorses use of the herb as a supportive treatment for chronic inflammatory liver conditions and cirrhosis. Average daily dose is 12–15 g of powdered seed.[107] This is equivalent to about 200–400 mg of silymarin. Unfortunately, silymarin is very poorly soluble in water, so the herb is not effective in the form of a tea. Studies

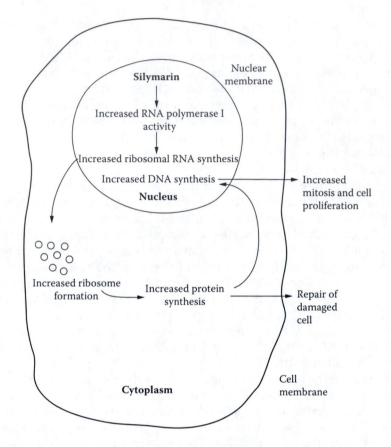

Figure 3.1 Mechanism of action of silymarin on the repair of the hepatocyte.

show that such a beverage contains less than 10 percent of the initial activity in the plant material.[108] Coupled with the fact that silymarin is relatively poorly absorbed (20–50 percent) from the GI tract, it is obvious that the administration of concentrated products is advantageous. Milk thistle is marketed in the United States in the form of capsules, usually containing 200 mg each of a concentrated extract representing 140 mg of silymarin. Toxic effects resulting from the consumption of milk thistle have apparently not been reported.

The only other liver-protective herb that need be mentioned at all, and then only briefly, follows.

Schizandra

Also spelled *schisandra,* this herb consists of the dried, ripe fruits of *Schisandra chinensis* (Turcz.) Baill. (family Schisandraceae). The fruits and seeds of this plant, native to China, have been extensively investigated chemically and found to contain more than thirty different lignans, including schizandrin, isoshizandrin, and gomisins A, B, C, D, F, and G. Some of these compounds have been studied in small animals for physiological activity. Several of them appear to protect the liver from toxic substances. Even though a number of schizandra products or combinations are currently marketed, their safety and efficacy for any purpose, particularly antihepatotoxic effects, remain unproven.[109] The consumption of schizandra cannot be recommended at this time.

Peptic ulcers

Peptic ulcer disease is a chronic inflammatory condition characterized by ulceration, most commonly in the first few centimeters of the duodenum (duodenal ulcers) and along the lesser curvature of the stomach (gastric ulcers). Peptic ulcer disease is typically a recurrent condition, with 50–90 percent of patients with duodenal ulcers having a recurrence within a year. In addition, nearly all patients with duodenal ulcer have gastritis caused by the bacterium *Helicobacter pylori,* which may account for the recurrent nature of the disease.

The goals in treatment of peptic ulcers are to relieve pain, enhance ulcer healing, and prevent recurrence. These are normally accomplished by neutralizing the hydrochloric acid of the stomach by the frequent administration of antacids, by decreasing the secretory activity of the stomach with the use of histamine H_2-receptor antagonists (cimetidine, ranitidine) or acid-pump inhibitors (omeprazole), by protecting the gastric mucosa with the administration of prostaglandins, and by preventing recurrence through the elimination of *H. pylori* with antibiotic therapy. Treatment should be intensive; during its course, the intake of irritants such as aspirin, other nonsteroidal anti-inflammatory agents, or alcohol must be strictly avoided.

Effective herbal treatments for ulcers are not numerous. Belladonna is a prescription drug and unavailable for self-treatment because of the potential toxicity of its alkaloidal constituents. Possibly, the only truly effective plant remedy is licorice.

Plant remedies for peptic ulcers

Licorice

Also known as glycyrrhiza, this herb consists of the dried rhizome (underground stem) and roots of *Glycyrrhiza glabra* L., known in commerce as Spanish licorice; of *G. glabra* L. var. *glandulifera* (Wald. & Kit.) Reg. & Herd., known in commerce as Russian licorice; or of other varieties of *G. glabra* L. that yield a yellow and sweet wood (family Fabaceae). It is often referred to as licorice root.

During World War II, a Dutch physician, F. E. Revers, noted that peptic ulcer patients in the small city of Heerenveen in the northern part of the Netherlands improved markedly when treated with a paste containing 40 percent licorice extract prepared by a local pharmacist.[110] Revers used licorice paste to treat a number of ulcer patients successfully, but in doing so he noted a serious side effect. About 20 percent of the patients developed edema, principally in the face and extremities. However, those unpleasant effects disappeared promptly and completely when treatment was discontinued.

During the intervening half century, a great many studies have been conducted on licorice, its constituents, and their effects. Much has been learned about the therapeutic usefulness and the side effects of the herb, but in essence, the findings remain the same. Licorice is useful in the treatment of peptic ulcers; depending on the dose, it may produce serious side effects. These undesirable effects are mineralocorticoid in nature. Specifically, they include headache, lethargy, sodium and water retention (the edema noted by Revers), excessive excretion of potassium, and high blood pressure. Eventually, heart failure or cardiac arrest may result. The medical literature is replete with references to cases of poisoning produced by overconsumption of licorice candy or licorice-containing tobacco.[111]

Glycyrrhizin (glycyrrhizic or glycyrrhizinic acid), a triterpene glycoside with saponin-like properties, is contained in licorice in a range of 2–14 percent. Plant material of good quality contains at least 4 percent. It is responsible for the sweet taste of the herb; it is some fifty times sweeter than sugar. The glycoside has pronounced expectorant and antitussive properties.[112] On hydrolysis, glycyrrhizin loses its sweet taste and is converted to glycyrrhetinic acid (glycyrrhetic acid) and two molecules of glucuronic acid. Both glycyrrhizin and its triterpene aglycone, glycyrrhetinic acid, possess distinct anti-inflammatory and antiallergic properties. These account for licorice's effectiveness in treating ulcers. Both compounds also possess mineralocorticoid activity that accounts for the herb's side effects.

Recent studies with glycyrrhetinic acid have revealed that the antiulcer effects of licorice are due to inhibition of 15-hydroxyprostaglandin dehydrogenase, an enzyme that metabolizes prostaglandins E_2 and $F_2\alpha$ to the inactive 15-ketoprostaglandins (see Figure 3.2). By blocking prostaglandin metabolism, the biological half-life of the active compounds is extended and has the effect of raising the local concentration of prostaglandins in the stomach, which promotes protective mucous secretions and cell proliferation of the gastric mucosa, leading to the healing of ulcers.[113]

15-hydroxyprostaglandin dehydrogenase belongs to a family of short-chain dehydrogenase reductase (SDR) enzymes that influence mammalian reproduction, hypertension, neoplasia, and digestion. Amino acid sequence analysis of enzymes in this family indicates that they have a

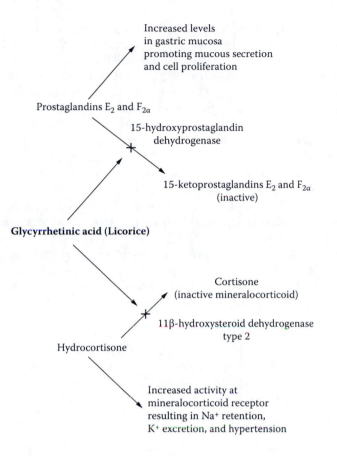

Figure 3.2 Glycyrrhetinic acid inhibition of short-chain dehydrogenase reductase (SDR) enzymes.

degree of sequence homology, and x-ray crystal structures of five members of the family demonstrate tertiary structure similarities.[114,115] Included in the SDR family is 11β-hydroxysteroid dehydrogenase, which is also inhibited by glycyrrhetinic acid.[113]

As illustrated in Figure 3.2, 11β-hydroxysteroid dehydrogenase acts as a "gatekeeper" and controls hormone access to the mineralocorticoid receptor. There are at least two isoforms of the enzyme. In the kidney, a high-affinity isoform designated type 2 converts the mineralocorticoid-active hydrocortisone to inactive cortisone, thereby protecting the mineralocorticoid receptor from hydrocortisone. Excessive licorice administration results in an inhibition of this conversion with a subsequent increase in hydrocortisone levels and the undesirable effects of mineralocorticoid activity: namely, sodium retention, potassium excretion, and high blood pressure.[116]

In summary, by inhibiting SDR enzymes, glycyrrhetinic acid produces a therapeutic effect by increasing prostaglandin levels and produces adverse effects by increasing hydrocortisone levels.

Carbenoxolone, a semisynthetic derivative of glycyrrhetinic acid, is widely marketed outside the United States as an antiulcer drug. It is not yet available in the United States, so persons wishing to utilize licorice for treatment of this condition are restricted to the herb itself. Normally, it is consumed in the form of a beverage prepared by adding about ½ cup of boiling water to 1 teaspoonful (2–4 g) of the herb and simmering the mixture for five minutes. After cooling, strain and drink this quantity of beverage three times daily after meals. The German Commission E has approved the use of licorice in ulcer therapy but cautions that the treatment should not be continued longer than four to six weeks.[117] It recommends a dosage level of 200–600 mg of glycyrrhizin daily; with an herb of average quality, this regimen would provide an amount about midpoint in that range. Elderly persons or those suffering from cardiovascular disease, liver or kidney problems, or potassium deficiency should avoid consuming licorice unless they do so while under the care of a physician.

Licorice also has a considerable reputation as an expectorant and cough suppressant, and it is frequently utilized in the treatment of symptoms associated with the common cold. Lozenges and candies containing licorice extract are especially suitable. However, particularly in the United States, one must make certain that they do contain real licorice. Most "licorice" candy manufactured in the United States is simply flavored with anise oil.

In addition to licorice's antiulcer and expectorant/cough suppressant activities, a number of other potential uses have been studied in small animals and, in some cases, in human beings.[118] As a result, the herb is postulated to possess hypolipidemic (cholesterol and triglyceride lowering), anticariogenic (antiplaque and anti-tooth-decay features), antimicrobial and antiviral, immunosuppressive, antianemia, and antihepatotoxic properties. Adequate evidence is not available to support the effectiveness of licorice in any of these conditions, but its local application as a hydrocortisone potentiator in preparations used to treat various skin conditions seems to hold considerable promise.[119]

Ginger
Investigations in small animals have shown that extracts of fresh ginger (see discussion in "Nausea and Vomiting") inhibited gastric secretion and the formation of stress-induced lesions.[119] The antiulcer activity of the herb requires confirmation but is certainly worthy of further study.

References

1. Tyler, V. E. *The new honest herbal,* 109–110. Philadelphia, PA: J. B. Lippincott.
2. Mowrey, D. B., and D. E. Clayson. 1982. *Lancet* I:655–657.
3. Stott, J. R. R., M. P. Hubble, and M. B. Spencer. 1985. Advisory Group for Aerospace Research and Development. *Conference Proceedings 372* 39:1–6.
4. Grøntved, A., and E. Hentzer. 1986. *ORL: Journal of Oto–Rhino–Laryngology and Its Related Specialties* 48:282–286.
5. Wood, C. D., J. E. Manno, M. J. Wood, B. R. Manno, and M. E. Mims. 1988. *Clinical Research Practices and Drug Regulatory Affairs* 6:129–136.
6. Grøntved, A., T. Brask, J. Kambskard, and E. Hentzer. 1988. *Acta Oto–Laryngologica* (Stockholm) 105:45–49.
7. Schmid, R., T. Schick, R. Steffen, A. Tschopp, and T. Wilk. 1994. *Journal of Travel Medicine* 1 (4): 203–206.
8. Riebenfeld, D., and L. Borzone. 1999. *Health Notes Review of Complementary and Integrative Medicine* 6:98–101.
9. Holtmann, S., A. H. Clarke, H. Scherer, and M. Höhn. 1989. *Acta Oto–Laryngologica* (Stockholm) 108:168–174.
10. Stewart, J. J., M. J. Wood, C. D. Wood, and M. M. Mims. 1991. *Pharmacology* 42:111–120.
11. Ernst, E., and M. H. Pittler. 2000. *British Journal of Anaesthesia* 84:367–371.
12. Arfeen, Z., H. Owen, J. L. Plummer, A. H. Ilsley, R. A. C. Sorby-Adams, and C. J. Doecke. 1995. *Anaesthesia and Intensive Care* 23:449–452.
13. Betz, O., P. Kranke, G. Geldner, H. Wulf, and L. H. J. Eberhart. 2005. *Forsch. Komplemetarmed. Klass. Naturheilkd.* 12 (1): 14–23.
14. *Bundesanzeiger* (Cologne, Germany): May 5, 1988.
15. Awang, D. V. C. 1992. *Canadian Pharmaceutical Journal* 125:309–311.
16. Kawai, T., K. Kinoshita, K. Koyama, and K. Takahashi. 1994. *Planta Medica* 60:17–20.
17. Phillips, S., R. Ruggier, and S. E. Hutchinson. 1993. *Anaesthesia* 48:715–717.
18. Visalyaputra, S., N. Petchpaisit, K. Sancharun, and R. Chouvaratna. 1998. *Anaesthesia* 53:506–510.
19. Bone, M. E., D. J. Wilkonson, J. R. Young, J. McNeil, and S. Charlont. 1990. *Anaesthesia* 45:669–671.
20. Fischer-Rasmussen, W., S. K. Kjaer, C. Dahl, and U. Asping. 1990. *European Journal of Obstetrics and Gynecological Reproductive Biology* 38:19–24.
21. Vutyavanich, T., T. Kraisarin, and R. A. Ruangsi. 2001. *Obstetrics and Gynecology* 97:577–582.
22. Backon, J. *Anaesthesia* 46:705–706.
23. Awang, D. V. C., and A. Fugh-Berman. 2002. *Alternative Therapies in Women's Health* 4 (8): 57–60.
24. Haas, H. *Arzneipflanzenkunde,* 78. Mannheim, Germany: B. I. Wissenschaftsverlag.
25. Hänsel, H. 1991. *Phytopharmaka,* 2nd ed., 121–123. Berlin: Springer–Verlag.
26. List, P. H., and L. Hörhammer, eds. 1973. *Hagers Handbuch der Pharmazeutischen Praxis,* 4th ed., vol. 4, 1115. Berlin: Springer–Verlag.
27. *Bundesanzeiger* (Cologne, Germany): November 30, 1985; March 6, 1990.
28. Tanaka, S., T. Furukawa, K. Adachi, and M. Ishimoto, M. 1988. *Iryo* 42:591–595.

29. Pahlow, M. 1985. *Das Grosse Buch der Heilpflanzen*, 330–331. Munich: Gräfe und Unzer.
30. *Bundesanzeiger* (Cologne, Germany): July 6, 1988; March 6, 1990.
31. Bisset, N. G., ed. 1994. *Herbal drugs and phytopharmaceuticals*, English ed. (*Teedrogen*, M. Wichtl, ed.), 45–48. Boca Raton, FL: CRC Press.
32. Brunton, L. L. 1996. Agents affecting gastrointestinal water influx and motility. In *Goodman and Gilman's the pharmacological basis of therapeutics*, 9th ed., ed. J. G. Hardman, L. E. Limbird, P. B. Molinoff, and R. W. Ruddon, 917–936. New York: McGraw–Hill.
33. Robbers, J. E., M. K. Speedie, and V. E. Tyler. 1996. *Pharmacognosy and pharmacobiotechnology*, 45–46. Baltimore, MD: Williams & Wilkins.
34. Ashraf, W., F. Park, J. Lof, and M. M. Quigley. 1995. *Alimentary Pharmacology and Therapeutics* 9:639–647.
35. Chesken, L. J., N. Kamal, M. D. Crowell, M. M. Schuster, and W. E. Whitehead. 1995. *Journal of the American Geriatrics Society* 43:666–669.
36. Anderson, J. W., N. Zettwoch, T. Feldman, J. Tietyen-Clark, P. Oeltgen, and C. W. Bishop. 1988. *Archives of Internal Medicine* 148:292–296.
37. Davidson, M. H., L. D. Dugan, J. H. Burns, D. Sugimoto, K. Story, and K. Drennan. 1996. *American Journal of Clinical Nutrition* 63:96–102.
38. Turley, S. D., B. P. Daggy, and J. M. Dietschy. 1996. *Journal of Cardiovascular Pharmacology* 27:71–79.
39. Alabaster O., Z. Tang, and N. Shivapurkar. 1996. *Mutation Research* 350:185–197.
40. Leng-Peschlow, E. 1993. *Pharmacology* 47 (suppl. 1): 14–21.
41. Robbers, J. E., M. K. Speedie, and V. E. Tyler. 1996. *Pharmacognosy and pharmacobiotechnology*, 50–52. Baltimore, MD: Williams & Wilkins.
42. Mereto, E., M. Ghia, and G. Brambilla. 1996. *Cancer Letters* 101:79–83.
43. Brusick, D., and U. Mengs. 1997. *Environmental and Molecular Mutagenesis* 29:1–9.
44. Robbers, J. E., M. K. Speedie, and V. E. Tyler. 1996. *Pharmacognosy and pharmacobiotechnology*, 52–53. Baltimore, MD: Williams & Wilkins.
45. Bisset, N. G., ed. 1994. *Herbal drugs and phytopharmaceuticals*, English ed. (*Teedrogen*, M. Wichtl, ed.), 208–211. Boca Raton, FL: CRC Press.
46. *Bundesanzeiger* (Cologne, Germany): December 5, 1984.
47. Jekat, F. W., H. Winterhoff, and F. H. Kemper. 1990. *Zeitschrift für Phytotherapie* 11:177–184.
48. Bisset, N. G., ed. 1994. *Herbal drugs and phytopharmaceuticals*, English ed. (*Teedrogen*, M. Wichtl, ed.), 463–466. Boca Raton, FL: CRC Press.
49. Beck, H., and K. Beck. 1982. *Schaufenster* (supplement to *Deutsche Apotheker Zeitung*) 31 (8/9): 33–34.
50. Hänsel, R., and T. Haas. 1984. *Therapie mit Phytopharmaka*, rev. ed., 137–140. Berlin: Springer–Verlag.
51. Bisset, N. G., ed. 1994. *Herbal drugs and phytopharmaceuticals*, English ed. (*Teedrogen*, M. Wichtl, ed.), 348–350. Boca Raton, FL: CRC Press.
52. Bisset, N. G., ed. 1994. *Herbal drugs and phytopharmaceuticals*, English ed. (*Teedrogen*, M. Wichtl, ed.), 431–433. Boca Raton, FL: CRC Press.
53. Reynolds, J. E. F., ed. 1996. *Martindale: The extra pharmacopoeia*, 31st ed., 1748. London: Royal Pharmaceutical Society of Great Britain.
54. *Bundesanzeiger* (Cologne, Germany): August 14, 1987.
55. *Bundesanzeiger* (Cologne, Germany): February 1, 1990.

56. Murray, M. T. 1995. *The healing power of herbs*, 2nd ed., 50–59. Rocklin, CA: Prima Publishing.
57. Schilcher, H. 1984. *Deutsche Apotheker Zeitung* 124:1433–1442.
58. Forster, H. B., H. Niklas, and S. Lutz. 1980. *Planta Medica* 40:309–319.
59. Hof-Mussler, S. 1990. *Deutsche Apotheker Zeitung* 130:2407–2410.
60. Robbers, J. E., M. K. Speedie, and V. E. Tyler. 1996. *Pharmacognosy and pharmacobiotechnology*, 87. Baltimore, MD: Williams & Wilkins.
61. Robbers, J. E., M. K. Speedie, and V. E. Tyler. 1996. *Pharmacognosy and pharmacobiotechnology*, 98–99. Baltimore, MD: Williams & Wilkins.
62. *Lawrence Review of Natural Products:* July 1990.
63. *Bundesanzeiger* (Cologne, Germany): November 30, 1985; March 13, 1986.
64. Blumenthal, M. 1990. *HerbalGram* 23:32–33, 49.
65. May, B., H.-D. Kuntz, M. Kieser, and S. Köhler. 1996. *Arzneimittel-Forschung* 46 (II): 1149–1153.
66. Beesley, A., J. Hardcastle, P. T. Hardcastle, and C. J. Taylor. 1996. *Gut* 39:214–219.
67. Wichtl, M. 1983. *Deutsche Apotheker Zeitung* 123:2114.
68. Wilkinson, S. M., and M. H. Beck. 1994. *Contact Dermatitis* 30:42.
69. Morton, C. A., J. Garioch, P. Todd, P. J. Lamey, and A. Forsyth. 1995. *Contact Dermatitis* 32:281–284.
70. Pahlow, M. 1985. *Das Grosse Buch der Heilpflanzen*, 258–260. Munich: Gräfe and Unzer.
71. Carle, R., and O. Isaac. 1987. *Zeitschrift für Phytotherapie* 8:67–77.
72. Schilcher, H. 1987. *Die Kamille*. Stuttgart: Wissenschaftliche Verlagsgesellschaft, 152 pp.
73. Mann, C., and J. Staba. 1986. The chemistry, pharmacology, and commercial formulations of chamomile. In *Herbs, spices, and medicinal plants: Recent advances in botany, horticulture, and pharmacology*, vol. 1, ed. L. E. Craker and J. E. Simon, 233–280. Phoenix, AZ: Oryx Press.
74. Bauer, R. 1992. *The American Herb Association Newsletter* 9 (1): 4.
75. Della Loggia, R., R. Carle, S. Sosa, and A. Tubaro. 1990. *Planta Medica* 56:657.
76. Lewis, W. H. 1992. *Economic Botany* 46:426–430.
77. Hausen, B. M., E. Busker, and R. Carle. 1984. *Planta Medica* 50:229–234.
78. Hausen, B. M. 1996. *American Journal of Contact Dermatitis* 7:94–99.
79. Awang, D. V. C. 2003. *Leung's (Chinese) Herb News* 40:2.
80. *Bundesanzeiger* (Cologne, Germany): December 5, 1984.
81. Carle, R., I. Fleischhauer, and D. Fehr. 1987. *Deutsche Apotheker Zeitung* 127:2451–2457.
82. Carle, R., I. Fleischauer, J. Beyer, and E. Reinhard. 1990. *Planta Medica* 56:456–460.
83. *Bundesanzeiger* (Cologne, Germany): July 6, 1988; February 1, 1990; November 30, 1985.
84. *Lawrence Review of Natural Products:* March 1996.
85. Hänsel, R. 1991. *Phytopharmaka*, 2nd ed., 186–191. Berlin: Springer–Verlag.
86. Ammon, H. P. T., and M. A. Wahl. 1991. *Planta Medica* 57:1–7.
87. Jaspersen-Schib, R. 1991. *Schweize Apotheker-Zeitung* 129:706–710.
88. Snow, J. M. 1995. *The Protocol Journal of Botanical Medicine* (Autumn): 43–46.
89. Soni, K. B., A. Rajan, and R. Kullan. 1992. *Cancer Letters* 66:115–121.
90. Soni, K. B., and R. Kuttan. 1992. *Indian Journal of Physiology and Pharmacology* 36:273–275.

91. Selvam, R., L. Subramanian, R. Gayathri, and N. Angayarkanni. 1995. *Journal of Ethnopharmacology* 47:59–67.
92. *Bundesanzeiger* (Cologne, Germany): November 30, 1985.
93. Bisset, N. G., ed. 1994. *Herbal drugs and phytopharmaceuticals,* English ed. (*Teedrogen,* M. Wichtl, ed.), 173–175. Boca Raton, FL: CRC Press.
94. *Lawrence Review of Natural Products:* May 1991.
95. *Bundesanzeiger* (Cologne, Germany): April 23, 1987.
96. Bisset, N. G., ed. 1994. *Herbal drugs and phytopharmaceuticals,* English ed. (*Teedrogen,* M. Wichtl, ed.), 486–489. Boca Raton, FL: CRC Press.
97. *Bundesanzeiger* (Cologne, Germany): December 5, 1984.
98. *Lawrence Review of Natural Products:* December 1987.
99. Dhiman, R. K., and Y. K. Chawla. 2005. *Digestive Disease and Sciences* 50 (10): 1807–1812.
100. Notka, F., G. Meier, and R. Wagner. 2004. *Antiviral Research* 64:93.
101. Huseini, H. F., S. M. Alavain, R. Heshmat, M. R. Heydari, and K. Abolmaali. 2005. *Phytomedicine* 12:619–624.
102. Wagner, H., and O. Seligmann. 1985. Liver therapeutic drugs from *Silybum marianum.* In *Advances in Chinese medicinal materials research,* ed. H. M. Chang, H. W. Yeung, W.-W. Tso, and A. Koo, 247–256. Singapore: World Scientific Publishing.
103. *Review of Natural Products:* January 1997.
104. Sonnenbichler, J., and I. Zetl. 1992. *Planta Medica* 58 (suppl.): A580.
105. Morazzoni, P., and E. Bombardelli. 1995. *Fitoterapia* 66:3–42.
106. Leng-Peschlow, E., and A. Strenge-Hesse. 1991. *Zeitschrift für Phytotherapie* 12:162–174.
107. *Bundesanzeiger* (Cologne, Germany): March 13, 1986.
108. Merfort, I., and G. Willuhn. 1985. *Deutsche Apotheker Zeitung* 125:695–696.
109. *Lawrence Review of Natural Products:* June 1988.
110. Nieman, C. 1962. *Chemist and Druggist* 177:741–745.
111. Tyler, V. E. 1987. *The new honest herbal,* 143–145. Philadelphia, PA: J. B. Lippincott.
112. Chandler, R. F. 1985. *Canadian Pharmaceutical Journal* 118:420–424.
113. Baker, M. E. 1994. *Steroids* 59:136–141.
114. Duax, W. L., J. F. Griffin, and D. Ghosh. 1996. *Current Opinion in Structural Biology* 6:813–823.
115. Duax, W. L., and D. Ghosh. 1997. *Steroids* 623:95–100.
116. Walker, B. R., and R. Best. 1995. *Endocrine Research* 21:379–387.
117. *Bundesanzeiger* (Cologne, Germany): May 15, 1986.
118. Wren, R. C. 1988. *Potter's new cyclopaedia of botanical drugs and preparations,* 173–175. Saffron Waldon, England: C. W. Daniel.
119. Teelucksingh, S., A. D. R. Mackie, D. Burt, M. A. McIntyre, L. Brett, and C. R. W. Edwards. 1990. *Lancet* 335:1060–1063.

chapter four

Kidney, urinary tract, and prostate problems

Infections and kidney stones

The kidneys function to remove waste from the body while preserving the chemical integrity of its cells and tissues. They work to maintain an appropriate concentration of electrolytes, amino acids, and glucose, as well as foreign substances, in the plasma and lymph. Although diuretics are commonly thought of as increasing the volume of urine excreted, they actually do much more than that.

There are now several types of widely used synthetic diuretics. These differ markedly in their exact site and mode of action, but basically all are used to treat various kinds of edema by enhancing both fluid and electrolyte excretion, thereby reducing the amount of extracellular fluid. Many of the beneficial effects of such diuretics, including the control of hypertension, derive not merely from their ability to stimulate the elimination of water but also of the electrolytes Na^+, Cl^-, and HCO_3^-.[1]

Technically, the classic herbal diuretic drugs are not diuretics at all but, rather, are more accurately designated *aquaretics*. Usually containing volatile oils, flavonoids, saponins, or tannins, they function to increase the volume of urine by promoting blood flow in the kidneys, thereby raising the glomerular filtration rate.[2] However, unlike the synthetic diuretics, they do not retard the resorption of Na^+ and Cl^- in the renal tubules, so quantities of these electrolytes are retained in the body and not excreted with the water. This means that the herbal aquaretics are not suited for the treatment of edema and especially not for hypertension.[3]

Nevertheless, these herbs can prove useful for certain other conditions—for example, minor infections—that benefit from an increased volume of urine. According to Schilcher, these include pyelonephritis (local infection of the renal tissues), urethritis or ureteritis (inflammation of the urethra or of a ureter), cystitis (inflammation of the urinary bladder), and the like.[4] Some of the herbs also exhibit antibacterial properties that, in combination with the increased urinary output, are useful in combating infection. Consequently, phytomedicinals with aquaretic or antiseptic properties will be considered here in the same category. In addition, one plant product is included primarily for its antiseptic properties and another because it prevents infection, rather than for either of these effects.

Another therapeutic application of an increased urine flow is in the prevention of kidney stones. Although there are many causes for the formation of kidney stones, the major determinant is an increased urinary concentration (supersaturation) of the constituents making up the stones such as Ca^{++}, oxalate, uric acid, and cystine. A low urine volume favors supersaturation of these constituents and can result in crystallization leading to the formation of stones.[5] Individuals with a history of stone formation are at risk for recurrence and should take measures to increase urine volume.

Significant aquaretic–antiseptic herbs

Goldenrod

Probably the most effective and safest of these herbs is goldenrod. Although its value is scarcely recognized in the United States, the aboveground parts of several species of *Solidago* are widely used in Europe to treat inflammations of the urinary tract and to prevent the formation or facilitate the elimination of kidney stones. In Europe, *S. virgaurea* L., *S. gigantea* Ait., *S. canadensis* L., and their hybrids are all used more or less interchangeably.

The genus *Solidago* of the family Asteraceae is notorious for its ability to hybridize, much to the consternation of botanists who recognize about 130 species in the United States.[6] Deam collected twenty-seven species in Indiana alone.[7] Although there are some qualitative and quantitative differences in their anti-inflammatory and bacteriostatic properties, the aquaretic properties of all the goldenrods that have been investigated are similar enough to permit them to be grouped together.[8] The information that follows is based largely on studies conducted on the European goldenrod (*S. virgaurea*).

Goldenrod contains a number of saponins based on polygalic acid; at least twelve diterpenes; phenolic glycosides, especially leiocarposide; and various miscellaneous flavonoids, tannins, polysaccharides, and the like.[9] The plant's aquaretic action is sometimes attributed to the flavonoids, but an isolated flavonoid mixture proved relatively inactive in animal studies.[10] Saponins may enhance their effects. The glycoside leiocarposide has been shown to have both anti-inflammatory and analgesic properties in tests carried out in rats.[11] These effects are probably due to the fact that the glycoside is hydrolyzed in the intestinal tract—although very slowly and incompletely—to salicylic acid.[12]

In spite of this relative metabolic stability of leiocarposide, Polish investigators noted that it had significant diuretic activity in rats.[13] Interestingly, they also found that the diuretic effect of the glycoside was reduced by the presence of flavonoids or saponins. The exact identity of the constituents responsible for the aquaretic effects of goldenrod thus remains somewhat controversial, although most authors continue to ascribe it to the contained mixture of flavonoids and saponins.[14]

More recently, a commercially available *S. gigantea* herb extract prepared using 60 percent ethanol as the extracting solvent revealed dose-related anti-inflammatory properties in the rat paw model. In high doses, these were almost as pronounced as the prescription drug diclofenac sodium. Further, a well-defined aquaretic and spasmolytic effect was observed.[15]

As is the case with almost all herbs, preliminary studies of goldenrod have reported other activities, including antifungal properties in the triterpene saponin fraction[16] and antitumor activity in the polysaccharides.[17] These are merely initial indications and have, as yet, no therapeutic utility.

In spite of some uncertainty regarding the nature of its active principles, a decoction of goldenrod is an effective aquaretic. It is prepared by adding 1–2 teaspoonfuls of the dried herb to ½ pint of water, bringing the water to a boil, and allowing it to stand for two minutes before straining and drinking.[18] The usual dose ranges from 0.2 to 0.4 oz of the herb per day.[19]

German Commission E has endorsed the herb to increase the amount of urine in inflammation of the kidney and bladder and as a preventive treatment in cases of kidney stones and gravel. Toxicity and contraindications are not reported. In this regard, it is superior to most other aquaretics of plant origin.

Parsley

All parts of the plant *Petroselinum crispum* (Mill.) A. W. Hill of the family Apiaceae contain varying amounts of a volatile oil with aquaretic activity; about 0.1 percent of the oil is found in the root, 0.3 percent in the leaf, and 2–7 percent in the fruit.[20,21] The aquaretic effect is attributed to the presence of two major components: myristicin and apiol. Concentrations of these constituents vary, depending on the variety of parsley from which the oil is obtained. However, some may contain 60–80 percent apiol and others 50–60 percent myristicin.

In addition to their aquaretic properties, both apiol and myristicin act as uterine stimulants, and the former was once widely used as an abortifacient drug. The oil also contains appreciable amounts of furanocoumarins or psoralens, compounds that cause photosensitivity on exposure to sunlight. Because of the presence in the volatile oil of these uterine-stimulant and photosensitizing constituents, use of the oil-rich seeds or the isolated oil, with their greater potential for toxicity, is not recommended.[22] The German Commission E does recommend the leaves and roots of parsley as aquaretics for irrigation in disorders of the urinary tract and therapy for the prevention and treatment of kidney gravel. Average daily dose is 6 g. Many times, this amount is consumed by persons who enjoy the Lebanese salad tabbouleh.

Nevertheless, for pregnant women and light-skinned persons who may be subject to phototoxic effects, the consumption of parsley, either as a nutrient or for therapeutic purposes, is best avoided. As is the case with other volatile-oil-containing aquaretics, parsley functions as an irritant to the epithelial tissues of the kidney, thus increasing the blood flow and glomerular filtration rate. For this reason, it should be used with considerable caution by those suffering from kidney disease.

Juniper

The most active but also the most potentially toxic of the volatile-oil-containing aquaretic–antiseptic herbs consists of the dried ripe fruits (berries) of *Juniperus communis* L. and its variety *depressa* Pursh of the family Cupressaceae. Quality fruits normally yield 1–2 percent of a volatile oil containing various terpene hydrocarbons, especially α- and β-pinenes; sesquiterpenes, such as carophyllene and cadinene; and its principal aquaretic, the terpene alcohol terpinen-4-ol.[23]

Both juniper berries and their steam-distilled oil have ancient reputations as diuretics and genitourinary antiseptics and were official in the first edition of the USP in 1820.[24] However, it was gradually recognized that, as a result of its irritant action, the drug caused injurious effects to the kidneys. It was deleted from the official compendia in 1960. Although the herb is recommended by German Commission E for the treatment of indigestion, it is not approved as a single-ingredient aquaretic.[25]

In traditional Swedish medicine, juniper has been used to treat wounds and inflammatory diseases. A recent study supports this therapeutic use by demonstrating that an aqueous extract of the berries has inhibitory activity in vitro on prostaglandin biosynthesis and platelet activating factor (PAF)-induced exocytosis.[26]

The exact composition of juniper oil is quite variable. Schilcher and colleagues have theorized that oils containing a low ratio (e.g., 3:1) of the irritating terpene hydrocarbons to the nonirritating, active aquaretic terpinen-4-ol do not exhibit nephrotoxicity. However, some oils have a hydrocarbon-to-alcohol ratio as high as 55:1 and are prone to cause kidney damage characterized by albuminuria or renal hematuria.[10] This hypothesis requires additional pharmacological testing.

More recently, Schilcher and Heil published a critical review of the literature from 1844 to 1993 and came to the conclusion that the nephrotoxic effects of juniper berries and oil have been confused with observations concerning the use of turpentine oil in veterinary medicine.[27] It is also possible that turpentine oil has been used to adulterate juniper oil, which would explain a high hydrocarbon-to-alcohol ratio. Further, they speculate that nephrotoxic effects have been erroneously suggested from pathological protein values in urine, which in fact are due to acute kidney infections rather than from the constituents of juniper.

Because there is no simple way for the potential consumer to measure the terpene hydrocarbon-to-alcohol ratio in the oil contained in juniper berries, the only sensible alternative is not to use the herb or the oil as a therapeutic agent. The very small amounts of the oil used to flavor gin are certainly no more harmful than the beverage alcohol itself.

Minor aquaretic herbs

Two minor herbal aquaretics require at least a brief mention. They are not as effective as the plants previously discussed and are used relatively seldom in the United States. However, they are common and modestly effective ingredients in many of the diuretic teas sold in Europe and have a long tradition of folkloric use there.

Birch leaves

These consist of the dried leaves of the silver or white birch, *Betula verrucosa* Erh. or *B. pubescens* Erh., family Betulaceae. They contain 2–3 percent of flavonoids, especially hyperoside and quercitrin, as well as various proanthocyanidins. Up to 0.5 percent of ascorbic acid and a trace of volatile oil are also present.[28]

Aqueous and alcoholic extracts of the leaves produced significant aquaresis in rats, an activity attributed to their flavonoid content.[29] The German Commission E has endorsed the use of birch leaves as a drug producing increased urinary output and of value in treating kidney and urinary tract infections.[30] Toxicity and contraindications have not been reported. An adequate intake of water by the patient is a necessity during such treatment.

Lovage root

The dried root of *Levisticum officinale* W. D. J. Koch, family Apiaceae, also possesses mild aquaretic properties. A heavy volatile oil occurs in the root, usually in amounts ranging from 0.6 to 1 percent. Up to 70 percent of the oil consists of alkylphthalides. Although these constituents are responsible for the characteristic odor of the oil, they play a limited role, if any, in its aquaretic activity.[31] That effect is due mainly to the terpene derivatives in the oil.

Small animal tests of lovage's aquaretic action have produced extremely variable results. In humans, its activity is probably less than that of juniper but greater than that produced by birch leaves. Two of the alkylphthalides present in the oil, butylidenaphthalide and ligustilide, possess antispasmodic properties.[32]

The German Commission E recognizes the efficacy of lovage root as an aquaretic for the treatment of urinary tract inflammation and the prevention of kidney stones.[33] Recommended daily dose is 4–8 g. Because, like all volatile-oil-containing diuretics, it induces aquaresis by irritating the kidney, the herb should not be used in cases of kidney disease. Further, furanocoumarins with photosensitizing properties are present in the oil, so persons using this remedy should avoid prolonged exposure to strong sunlight.

Antiseptic herbs

Bearberry

The most effective antibacterial herb for urinary tract infections, bearberry, is often characterized as a diuretic, but, in fact, its aquaretic properties are minimal. Consisting of the dried leaves of *Arctostaphylos uva-ursi* (L.) Spreng, this member of the family Ericaceae is represented in the United States by two principal varieties: *coactylis* and *adenotricha* Fern. & Macbr. For many years, it was widely used as a urinary antiseptic, as indicated by its status in the official compendia (USP and NF) from 1820 to 1950.[34] The cessation of its widespread use coincided with the development of sulfa drugs and antibiotics that proved very effective against urinary tract infections.

Bearberry (or uva ursi leaves, as they are commonly called) contains a number of constituents including flavonoids, tannins, organic acids, and the like; however, antiseptic properties are due to the presence of two phenolic glycosides—arbutin and methylarbutin—present in concentrations ranging from 5 to 15 percent.[35] When the plant or an extract of it is consumed, arbutin is hydrolyzed in the intestinal tract to yield hydroquinone. Following absorption, this compound is bound as glycuronides and sulfate esters that are excreted in the urine. If the urine is alkaline (greater than pH 8), the conjugates, especially the sulfate esters, are partially saponified, and the hydroquinone thus freed produces an antibacterial effect.[36]

Obviously, bearberry will be most effective if the urine is maintained at an alkaline pH. This requires consumption of a diet rich in milk, vegetables (especially tomatoes), fruits, fruit juices, potatoes, etc. In addition, consumption of 6–8 g of sodium bicarbonate per day will ensure alkalinity. Because it is impractical to maintain an alkaline urine for an extended period of time, the utility of bearberry as a urinary antiseptic is greatly reduced.

The German Commission E recommends bearberry as a treatment for inflammatory conditions of the urinary tract; it is considered to be a bacteriostatic, rather than an aquaretic.[37] The usual dose is 10 g (1/3 oz) daily, corresponding to approximately 400–700 mg of arbutin. Because the leaves are rich in tannin, a suitable beverage is best prepared by soaking the leaves in a quantity of cold water overnight. In this way, much less tannin is dissolved compared to a tea prepared with boiling water. Because of the potential toxicity of hydroquinone and the impracticality of maintaining an alkaline urine, bearberry should be utilized only for relatively short periods of time—a few days at most.

Anti-infective herbs

Cranberry

One of the most useful herbs for the prevention and treatment of urinary tract infections (UTIs), cranberry, lacks any antiseptic or antibiotic properties per se. The sweetened, diluted juice of the American cranberry, *Vaccinium macrocarpon* Ait. (family Ericaceae)—also known as the trailing swamp cranberry—is marketed as cranberry juice cocktail. It was reported in 1923 that the urine of test subjects became more acid after eating large amounts of cranberries.[38] Because an acid medium hinders bacterial development, it was postulated that the berries might be useful in preventing or curing UTI, a condition especially prevalent among women. At the time, conventional medical treatments were largely ineffective.

In consequence, many women suffering from this condition began to consume quantities of the cocktail and reported good results. Such word-of-mouth recommendations were supplemented occasionally by articles in regional medical journals. One of the latter reported symptomatic relief from chronic kidney inflammation in female patients who drank 6 oz of cranberry juice twice daily. Even though a 1967 study showed that consumption of the commercial cranberry juice cocktail did not appreciably acidify the urine of consumers,[39] UTI sufferers continued to drink it and to report beneficial results.

It is now recognized that the effectiveness of cranberry juice in treating UTI results not from its acidifying properties but, rather, from its ability to prevent the microorganisms from adhering to the epithelial cells that line the urinary tract.[40,41] The most common of the UTI-causing bacteria is *Escherichia coli*. Many Gram-negative bacteria, such as *E. coli*, have thin, hairlike appendages covering their surface that are called fimbriae or pili. These fimbriae enable the bacteria to adhere to various epithelial surfaces for the purpose of colonization and are therefore referred to as adhesins or colonization factors. *E. coli* can make at least two types of adhesins. Most strains produce type 1 fimbriae, which adhere to epithelial cell receptors containing d-mannose.

Most uropathogenic *E. coli* strains that cause urinary tract infections commonly do not have d-mannose-mediated adherence but, rather, have P fimbrial adhesions, which attach to a portion of the P blood group antigen and recognize the disaccharide α-d-galactopyranoside-(1→4)-β-d-galactopyranoside (GAL–GAL). As illustrated in Figure 4.1, the P fimbrial adhesin mediates the adherence of the *E. coli* to uroepithelial cells that contain the GAL–GAL recognition site. Upon adherence, the bacterial cells multiply rapidly, causing infection.[41,42]

E. coli adhesin activity is inhibited by two different constituents of cranberry juice. One of these is the nearly ubiquitous fructose, which is found in all fruit juices and has limited significance in the prevention of

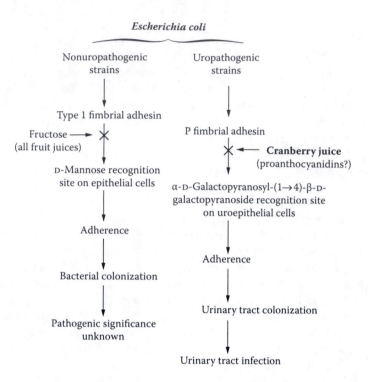

Figure 4.1 The mechanism of action of cranberry juice in preventing *Escherichia coli* urinary tract infection.

UTI. Fructose inhibits type 1 fimbrial adhesion, which does not have a role in the establishment of *E. coli* UTI. The other constituent is a high-molecular-weight compound that is incompletely identified. It is not found in other fruit juices (with the exception of blueberry juice) and acts specifically to inhibit P fimbrial adhesin that is expressed by uropathogenic strains of *E. coli* by binding to the bacterial surface, possibly to the adhesin itself (see Figure 4.1). Purification of this high-molecular-weight inhibitor from 1.5 L of cranberry juice cocktail using dialysis and fractionation on polyacrylamide resin resulted in 100 mg of an active fraction comprising mainly condensed tannins (proanthocyanidins).[43] Other constituents in cranberry include various carbohydrates and fiber, as well as a number of plant acids, including benzoic, citric, malic, and quinic.[44]

Recent clinical studies that support the efficacy of cranberry juice in the prevention and treatment of UTIs include a randomized, double-blind, placebo-controlled trial to measure the effect of regular ingestion of cranberry juice on the bacterial flora of older women. The study sample consisted of 153 women who were randomly allocated to an experimental group that drank 300 mL per day of a standard commercially available cranberry beverage containing 26 percent juice. The controls were given a synthetic placebo drink, which had an identical taste and color but contained no cranberry product. Urine samples were collected monthly for six months and tested for the presence of bacteria and white blood cells. It was found that women assigned to the cranberry beverage group had significantly less bacteriuria with pyuria than controls. The pH measurements of urines in both groups were nearly identical, indicating that urinary acidification did not explain the findings.[45]

In the Netherlands, using a randomized controlled crossover design, one group of patients in a nursing department of a general hospital was given 15 mL of cranberry juice mixed with water twice a day, and the control group received the same volume of water each day. After four weeks, the regimens were changed between groups, and a urine culture was taken after each four-week period. It was found that there were fewer cases of bacteriuria in patients during the period when they were administered cranberry juice.[46]

Recommended dosage of cranberry juice cocktail as a UTI preventive is 3 fluid oz (about one-third of which is pure juice); as a UTI treatment, consumption should be increased into the range of 12–32 fluid oz daily. An artificially sweetened product is available if sugar intake is to be limited.

Capsules containing dried cranberry powder are also available; six are said to be equivalent to 3 fluid oz of cocktail. Fresh or frozen cranberries may also be consumed; 1.5 oz is equivalent to 3 fluid oz of cocktail. In practice, this is scarcely feasible because of the high acidity and extremely sour taste of the raw berries.

Caution: Cranberry concentrate tablets available at nutrition stores are promoted for urinary tract health and are popular among women for preventing bladder infections. However, due to cranberry's high concentration of oxalate, which contributes to the development of kidney stones (nephrolithiasis), cranberry concentrates should be consumed with caution. In response to a case of acute nephrolithiasis, researchers at Stanford University Medical Center subjected five healthy female volunteers to the manufacturer's recommended dosage of cranberry tablets for seven days. At the onset of treatment and at the end of the seven-day period, twenty-four-hour urine collections were tested for pH, volume creatinine, oxalate, calcium, phosphate, uric acid sodium, citrate, magnesium and potassium, Urinary oxalate levels in the volunteers were significantly increased ($p = 0.01$) by an average of 43.4 percent, and excretion of potential lithogenic ions calcium, phosphate, and sodium also increased. The authors of the study advise that "physicians and manufacturers of cranberry products should make an effort to educate patients at risk for nephrolithiasis against ingestion of these dietary supplements."[47]

Cranberry–warfarin interaction?

In September 2003, the UK Committee on Safety of Medicines (CSM) issued a warning: "Patients taking warfarin should limit or avoid drinking cranberry juice."[48] No such warning has emerged in the United States. The UK judgment was based on five case study reports (one involving a death[49]) and speculated that cranberry juice inhibited the primary cytochrome P450 isoenzyme that metabolizes warfarin, CYP 2C9.

On December 23, 2003, the Cranberry Institute in the United States issued a press release that considered the UK advisory unwarranted, based on the opinion of experts whom they had engaged to review the evidence, which was viewed as inadequate. Nonetheless, the institute encouraged patients to consult with their physicians, promising to monitor emerging reports and support further research on warfarin interactions.[50] In early 2004, another case report concerned a patient with a heart valve replacement who exhibited elevated INR (international normalized ratio) clotting time and bleeding after commencing to drink cranberry juice cocktail on a regular basis.[51]

In addition to the theory of CYP 2C9 inhibition, other theories advanced to explain a potential warfarin interaction have included direct anticoagulant or antiplatelet effects, inhibition of vitamin K absorption, or inhibiting bacterial adhesion effecting changes to the microbial population of the large intestine and causing a shift away from bacteria that produce vitamin K.[52] None of these theories has been established thus far.

Benign prostatic hyperplasia (prostate enlargement)

The majority of men who consult urologists do so because of some kind of impairment of the urinary flow. In the group under forty-five years of age, this is usually due to prostatitis, which is an inflammation—frequently an infection—of the prostate gland. In men over forty-five, the cause is often benign prostatic hyperplasia (BPH) brought about by an abnormal, but nonmalignant, proliferation of cells and tissues of the gland. Histologic evidence of BPH is found in more than 50 percent of sixty-year-old men and in 90 percent of men by age eighty-five. Eventually, urethral obstruction leads to urinary retention, kidney damage, and infection. In advanced stages, surgical resection is the treatment of choice. In 1990, more than three hundred thousand operations were performed for BPH in the United States at a cost of more than two billion dollars.

To understand how medications may affect BPH, it is necessary to understand the mechanism that causes the condition. In the prostate, as is the case with many accessory sex organs, testosterone from the blood penetrates the prostatic cell by passive diffusion and is converted by the enzyme steroid 5α-reductase to the more potent androgen, dihydrotestosterone (DHT). The DHT binds to a specific receptor in the cytoplasm and this DHT-receptor complex is transported to the nucleus, where it initiates ribonucleic acid (RNA) and deoxyribonucleic acid (DNA) synthesis. This, in turn, results in protein synthesis, cell metabolism, and cell division.

In the normal growth process, when sex accessory organs reach a certain size, their further development is no longer influenced by testosterone or DHT. However, four to six times the normal amount of DHT is found in the hyperplastic prostate. Apparently, this high concentration of the hormone results in increased growth of the gland in mature males, but the reason for its presence and the manner in which it leads to hyperplasia are poorly understood.

In clinically significant BPH (enlarged prostate with clinical symptoms), hyperplastic nodules develop in the periurethral zone of the prostate gland. As these nodules enlarge, they compress both the surrounding prostatic tissue and the urethra, leading to a mechanical obstruction of urine flow from the bladder. In addition to the mechanical obstruction, the smooth muscle of the prostate capsule and the proximal urethra, as well as the trigone muscle of the bladder, are rich in α_1-adrenergic receptors that create a dynamic component of obstruction that fluctuates with autonomic stimulation.

The most common treatment for BPH is partial prostatectomy; however, in the past few years interest in nonsurgical therapy, including the use of drugs, has increased. Drug therapy can be considered in patients who do not have absolute indication for surgery or who are poor surgical risks, but at the present time it is generally not considered an alternative to surgery.

A second drug therapy approach is to use selective α_1-receptor blockers such as terazosin (Hytrin). Blocking the α_1-adrenergic receptors relaxes the smooth muscles of the prostate, thereby decreasing tone and increasing urinary flow rates (Figure 4.2). These drugs were originally developed for treating hypertension. Side effects occurring in 10–15 percent of patients include orthostatic hypotension, dizziness, tiredness, and headache.[53]

Although herbal medicines are widely used to treat the early stages of BPH in Europe, it must be noted that OTC drug products that claim to treat BPH cannot be sold in the United States. The FDA has banned their sale for two reasons: (1) the agency has not received evidence proving their effectiveness, and (2) their use may delay proper medical treatment.[54]

Substantial clinical evidence has now accumulated that makes the first reason for the FDA ban invalid, and two herbal medicines have received approval by German health authorities.[55] The second reason remains valid, and patients who have the obstructive and irritative symptoms of BPH should seek physician care because these symptoms may be caused

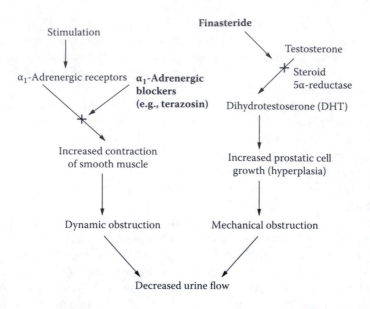

Figure 4.2 The mechanism of drug therapy in the treatment of benign prostatic hyperplasia (BPH).

by other conditions such as prostatic and bladder carcinomas, neurogenic bladder, and urinary tract infection.

Herbal remedies for benign prostatic hyperplasia (BHP)

Saw palmetto (sabal)

Saw palmetto consists of the partially dried, ripe fruit of *Serenoa repens* (Bartr.) Small (synonym *Sabal serrulata* Schult.f.), a low, scrubby palm of the family Arecaceae that grows from South Carolina south to Florida. The name sabal, although frequently applied to this small fan palm, is probably best reserved for the species of usually tall, tree-like palms belonging to the genus *Sabal*. Under the title Serenoa, saw palmetto was an "official" drug from 1906 to 1950 and was once widely used for a variety of ailments, particularly those of the urogenital type, before falling into near-oblivion in the United States after World War II.

European scientists, however, continued to study saw palmetto and recognized that, in patients suffering from BPH, an extract of the fruits produced increased urinary flow, reduced residual urine, increased ease in commencing micturition, and decreased frequency of urination.[56] However, the mechanism of action responsible for this beneficial activity is not clear. It has recently been demonstrated in an in vitro study that an ethanolic extract of saw palmetto inhibited dose dependently steroid

5α-reductase activity in the epithelium and stroma of human BPH; the mean inhibition was 29 and 45 percent, respectively.[57]

This would suggest a mechanism similar to finasteride; however, earlier in vitro comparative studies indicated that saw palmetto extract is several thousand times less potent as a 5α-reductase inhibitor than finasteride, and the concentration needed to produce inhibition is higher than that used in clinical therapy.[58] In addition, steroid 5α-reductase inhibition should result in a decrease in prostate size (see Figure 4.2), and this has not been demonstrated with saw palmetto treatment.[59]

The relief of symptoms without a significant decrease in prostate size suggests that saw palmetto extracts act on the dynamic obstruction by diminishing prostatic smooth muscle tone. Evidence has been obtained that saw palmetto produces a spasmolytic effect on smooth muscle contraction by inhibition of calcium influx at the plasma membrane level[60] rather than by blocking α_1-adrenergic receptors. Presumably, activation of the Na^+/Ca^{++} exchanger, perturbation of intracellular calcium mobilization, possibly mediated by cAMP, and posttranscriptional induction of protein synthesis might be important components of the spasmolytic action.[61]

In addition to their antiandrogenic and spasmolytic properties, anti-inflammatory or antiedematous activity has been demonstrated in the berries. This apparently results from inhibition of the cyclooxygenase and 5-lipoxygenase pathways, thereby preventing the biosynthesis of inflammation-producing prostaglandins and leukotrienes. The antiedematous activity in all likelihood is caused by inhibition of the arachidonic acid cascade.[62]

Together, the antiandrogenic, spasmolytic, and anti-inflammatory effects seem to account for the beneficial role of the herb in treating BPH. Placebo-controlled, double-blind clinical studies carried out on more than two thousand BPH patients in Germany have confirmed the effectiveness of a saw palmetto extract in such conditions.[63] In a three-month open trial, 505 patients with mild to moderate symptoms of BPH were treated with saw palmetto extract. The efficacy of the treatment was evaluated in 305 patients and symptoms were significantly improved after only forty-five days. After ninety days of treatment, 88 percent of patients and 88 percent of treating physicians considered the therapy effective. The incidence of side effects was 5 percent and compares favorably with existing drug therapy.[64] A recent comparative study between saw palmetto extract and synthetic agents (finasteride and α_1-adrenergic blocking agents alfuzosin and terazosin) indicated that saw palmetto extract had a higher benefit-to-risk ratio and a lower cost to patients than the synthetic agents.[65]

Chemical examination of saw palmetto has identified a relatively large number of constituents of the contained volatile and fatty oils, including, in the latter, large amounts of β-sitosterol-3-d-glucoside. Various acids, such as anthranilic, caffeic, and chlorogenic, as well as tannin, sugars, and polysaccharides are also present.[66] Unfortunately, the active antiandrogenic principles remain unidentified, but they are known to reside in the acidic lipophilic fraction of the drupes. A water-soluble polysaccharide was once reported to possess anti-inflammatory properties, but a more recent study found that the polysaccharides, β-sitosterol derivatives, and flavonoids all lacked anti-inflammatory effects when given orally.[67]

It thus appears that the principal activity of the fruit resides in the nonpolar constituents. In this regard, it was found recently that in a steroid 5α-reductase in vitro assay using human BPH tissue obtained by suprapubic prostatectomy, lauric acid and myristic acid were the major active constituents in an ethanolic extract of saw palmetto fruits.[57] These results correlate with the finding that brine shrimp lethality-directed fractionation of the 95 percent ethanol extract of berries of saw palmetto led to the isolation of the monoacylglycerides of lauric acid and myristic acid.[68]

German health authorities confirm the position that the active constituent is nonpolar by specifying a daily dose of saw palmetto of 1–2 g or 320 mg of an extract prepared by extracting the drug with a lipophilic solvent such as hexane or 90 percent alcohol. Although an aqueous extract of the berries possesses antiexudative properties, such action, if present, is minimal in comparison to that of the lipophilic extract.

This has important implications for those intending to use the crude drug in the customary form of a tea. Obviously, an aqueous beverage prepared from it would not contain the water-insoluble active constituents, so a preparation of this kind would have little value. For best effect, either the whole drug or an extract prepared with a nonpolar solvent must be used.

Nettle root

Use of the root of the stinging nettle, *Urtica dioica* L.; the small stinging nettle, *U. urens* L.; or hybrids of these members of the family Urticaceae for treatment of BPH is a relatively recent innovation in phytomedicine. A number of clinical studies support the plant's effectiveness. Chemical analysis of nettle root has resulted in the isolation and identification of a number of low- and high-molecular-weight compounds. The former include various lignans, scopoletin, sitosterol, sitosterol-3-*O*-glucoside, oleanolic acid, and 9-hydroxyoctadeca-10-*trans*,12-*cis*-dienoic acid. High-molecular-weight compounds include isolectins and five acid and neutral polysaccharides. The identity of the active principle and consequently its mechanism of action remain unknown.[69]

It has been postulated that the herb may have an effect on the amount of free (active) testosterone circulating in the blood or that it may inhibit aromatase, one of the key enzymes responsible for testosterone synthesis. Another, more recent theory attributes the activity to the presence of a lectin (protein) mixture designated UDA (*Urtica dioica* agglutinin) and several polysaccharides. UDA is unusually stable to acids and heat; consequently, it would retain its activity on oral administration.[70] In addition, the aqueous extract of the root has the ability to inhibit binding of sex hormone-binding globulin (SHBG) to receptors on prostatic membranes. It is postulated that the binding of an appropriate steroid with this globulin is involved in prostatic growth.[71] None of these various postulates regarding nettle root activity has been conclusively proven.

German health authorities have concluded that nettle root is an effective treatment for urinary difficulties arising from the early stages of prostate adenoma or BPH.[72] The usual dose is 4–6 g daily. Because the active principles are apparently water soluble, the root may be administered in the form of a tea. Contraindications are unknown, and side effects, consisting mostly of gastrointestinal disturbances, are minimal.

The dried leaves of the nettle plant are commonly employed as an aquaretic, and their consumption does result in an increase in the flow of urine.[73] The active principles responsible for this effect have not been identified. Nettle leaves are ordinarily taken in the form of a tea prepared from 3–4 teaspoonfuls (about 4 g) of the botanical and 150 mL of boiling water. One cup may be drunk three to four times daily together with additional water. As is the case with other herbal aquaretics, nettle leaf is not effective for hypertension or for edema resulting from cardiac insufficiency.

Pygeum

The lipophilic extract of the bark of *Prunus africana* (Hook f.) Kalkm. (synonym *Pygeum africanum* Hook f.) (family Rosaceae), an evergreen tree native to southern and central Africa, has been used in symptomatic treatment of mild to moderate BPH in Europe since 1969.[74]

The extract contains at least three classes of active constituents that exert a beneficial effect on BPH. Phytosterols present in both free and conjugated form compete with androgen precursors and also inhibit prostaglandin biosynthesis. Pentacyclic terpenes, including oleanolic, crataegolic, and ursolic acids, exhibit anti-inflammatory activity by inhibiting the enzymes involved in the depolymerization of proteoglycans in the connective tissue. Ferulic acid esters of fatty alcohols reduce the level of cholesterol in the prostate, causing a decrease in precursor for androgen synthesis.

In general, the results of twenty-six clinical trials conducted over the past twenty years involving more than six hundred patients indicate that administration of 100–200 mg daily of the lipophilic extract will produce significant improvement in the symptoms of BPH. The extract appears to be devoid of serious side effects, and it is well tolerated in humans during long-term administration.[75]

References

1. Berndt, W. O., and R. E. Stitzel. 1994. Water, electrolyte metabolism, and diuretic drugs. In *Modern pharmacology*, 4th ed., ed. C. R. Craig and R. E. Stitzel, 211–228. Boston: Little, Brown.

2. Schilcher, H., and D. Emmrich. 1992. *Deutsche Apotheker Zeitung* 132:2549–2555.
3. Hänsel, R., and H. Haas. 1984. *Therapie mit Phytopharmaka*, 206. Berlin: Springer–Verlag.
4. Schilcher, H. 1991. *Deutsche Apotheker Zeitung* 131:838–840.
5. Cotran, R. S., V. Kumar, and S. L. Robbins. 1994. *Robbins' pathologic basis of disease*, 5th ed., 984–985. Philadelphia, PA: W. B. Saunders and Company.
6. Begg, V. L. 1991. *Herb Quarterly* 50:33–35.
7. Deam, C. C. 1940. *Flora of Indiana*, 919–928. Indianapolis: State of Indiana Department of Conservation, Division of Forestry.
8. Hiller, K., and G. Bader. 1996. *Zeitschrift für Phytotherapie* 17:123–130.
9. Bisset, N. G., ed. 1994. *Herbal drugs and phytopharmaceuticals*, English ed. (*Teedrogen*, M. Wichtl, ed.), 530–533. Boca Raton, FL: CRC Press.
10. Schilcher, H., R. Boesel, S. Effenberger, and S. Segebrecht. 1989. *Zeitschrift für Phytotherapie* 10:77–82.
11. Metzner, J., R. Hirschelmann, and K. Hiller. 1984. *Pharmazie* 39:869–870.
12. Foetsch, G., S. Pfeifer, M. Bartoszek, P. Franke, and K. Hiller. 1989. *Pharmazie* 44:555–558.
13. Chodera, A., K. Dabrowska, M. Senczak, A. Wasik-Olejnik, L. Skrzypczak, J. Budzianowski, and M. Ellnain-Wojtaszek. 1985. *Acta Poloniae Pharmaceutica* 42:199–204.
14. Reznicek, G., J. Jurenitsch, M. Freiler, S. Korhammer, E. Haslinger, K. Hiller, and W. Kubelka. 1992. *Planta Medica* 58:94–98.
15. Leuschner, J. 1995. *Arzneimittel Forschung* 45:165–168.
16. Bader, G., K. Binder, K. Hiller, and H. Ziegler-Böhme. 1987. *Pharmazie* 42:140.
17. Kraus, J., M. Schneider, and G. Franz. 1986. *Deutsche Apotheker Zeitung* 126:2045–2049.
18. Pahlow, M. 1979. *Das Grosse Buch der Heilpflanzen*, 147. Munich: Gräfe und Unzer.
19. *Bundesanzeiger* (Cologne, Germany): April 14, 1987; March 6, 1990.
20. *Lawrence Review of Natural Products:* February 1991.
21. Warncke, D. 1994. *Zeitschrift für Phytotherapie* 15:50–58.
22. *Bundesanzeiger* (Cologne, Germany): January 5, 1989.
23. Steinegger, E., and R. Hänsel. 1988. *Lehrbuch der Pharmakognosie und Phytopharmazie*, 4th ed., 319–321. Berlin: Springer–Verlag.
24. Claus, E. P. 1956. *Gathercoal and wirth pharmacognosy*, 3rd ed., 290–291. Philadelphia, PA: Lea & Febiger.
25. *Bundesanzeiger* (Cologne, Germany): December 5, 1984.
26. Tunón, H., C. Olavsdotter, and L. Bohlin. 1995. *Journal of Ethnopharmacology* 48:61–76.
27. Schilcher, H., and B. M. Heil. 1994. *Zeitschrift für Phytotherapie* 15:205–213.
28. Steinegger, E., and R. Hänsel. 1988. *Lehrbuch der Pharmakognosie und Phytopharmazie*, 4th ed., 564–565. Berlin: Springer–Verlag.
29. Schilcher, H., and H. Rau. 1988. *Urologe B* 28:274–280.
30. *Bundesanzeiger* (Cologne, Germany): March 13, 1986.
31. Vollmann, C. 1988. *Zeitschrift für Phytotherapie* 9:128–132.
32. Bisset, N. G., ed. 1994. *Herbal drugs and phytopharmaceuticals*, English ed. (*Teedrogen*, M. Wichtl, ed.), 295–297. Boca Raton, FL: CRC Press.
33. *Bundesanzeiger* (Cologne, Germany): June 1, 1990.

34. Claus, E. P., and V. E. Tyler, Jr. 1965. *Pharmacognosy,* 5th ed., 152–153. Philadelphia, PA: Lea & Fibiger.
35. *Review of Natural Products:* July 1997.
36. Steinegger, E., and R. Hänsel. 1988. *Lehrbuch der Pharmakognosie und Phytopharmazie,* 4th ed., 696–699. Berlin: Springer–Verlag.
37. *Bundesanzeiger* (Cologne, Germany): December 5, 1984.
38. Blatherwick, N. R., and M. L. Long. 1923. *Journal of Biological Chemistry* 57:815–818.
39. *Lawrence Review of Natural Products:* July 1994.
40. Sabota, A. E. 1984. *Journal of Urology* 131:1013–1016.
41. Soloway, M. S., and R. A. Smith. 1988. *Journal of the American Medical Association* 260:1465.
42. Ofek, I., J. Goldhar, D. Zafriri, H. Lis, and N. Sharon. 1991. *New England Journal of Medicine* 324:1599.
43. Ofek, I., J. Goldhar, and N. Sharon. 1996. *Advances in Experimental Medicine and Biology* 408:179–183.
44. Hughes, B. G., and L. D. Lawson. 1989. *American Journal of Hospital Pharmacy* 46:1129.
45. Avorn, J., M. Monane, J. H. Gurwitz, R. J. Glynn, I. Choodnovskiy, and L. A. Lipsitz. 1994. *Journal of the American Medical Association* 271:751–754.
46. Haverkorn, M. J., and J. Mandigers. 1994. *Journal of the American Medical Association* 272:590.
47. Terris, M. K., M. M. Issa, and J. R. Tacker. 2001. *Urology* 57 (1): 26–29.
48. Committee on Safety Medicines. 2003. *Current Problems in Pharmacovigilance* 29:8.
49. Suvarna, R., M. Pirmohammed, and L. Henderson. 2003. *British Medical Journal* 327:1454.
50. www.Cranberryinstitute.org
51. Grant, P. 2004. *Journal of Heart Valve Disease* 13:25–26.
52. Howell, A. B. 2002. *Critical Reviews of Food Science and Nutrition* 42 (3 suppl.): 273–278.
53. Isacksen, R. R., and J. S. Wheeler, Jr. 1997. Benign prostatic hyperplasia. In *Conn's current therapy,* ed. R. E. Rakel, 689–693. Philadelphia, PA: W. B. Saunders Company.
54. 1990. *American Pharmacy NS* 30:321.
55. *Bundesanzeiger* (Cologne, Germany): January 5, 1989; February 1, 1990; March 6, 1990.
56. Engelman, U. 1997. *Zeitschrift für Phytotherapie* 18:13–19.
57. Weiser, H., S. Tunn, B. Behnke, and M. Krieg. 1996. *Prostate* 28:300–306.
58. Rhodes, L., R. L. Primka, C. Berman, G. Verguit, M. Gabriel, M. Pierre-Malice, and B. Gibelin. 1993. *Prostate* 22:43–51.
59. Vahlensieck, W., A. Volp, W. Lubos, and M. Kuntze. 1993. *Fortschritte der Medizin* 111:323–326.
60. Gutierrez, M., M. J. Garcia de Boto, B. Cantabrana, and A. Hidalgo. 1996. *General Pharmacology* 27:171–176.
61. Gutierrez, M., A. Hidalgo, and B. Cantabrana. 1996. *Planta Medica* 623:507–511.
62. Breu, W., M. Hagenlocher, K. Redl, G. Tittel, F. Stadler, and H. Wagner. 1992. *Arzneimittel Forschung* 42:547–551.

63. Breu, W., F. Stadler, M. Hagenlocher, and H. Wagner. 1992. *Zeitschrift für Phytotherapie* 13:107–115.
64. Braeckman, J. 1994. *Current Therapeutic Research* 55:776–785.
65. Bach, D., M. Schmitt, and L. Ebeling. 1996. *Phytomedicine* 3:309–313.
66. Hänsel, R., and H. Haas. 1984. *Therapie mit Phytopharmaka*, 202. Berlin: Springer–Verlag.
67. Hiermann, A. 1989. *Archiv der Pharmazie* 322:111–114.
68. Shimada, H., V. E. Tyler, and J. L. McLaughlin. 1997. *Journal of Natural Products* 60:417–418.
69. Wagner, H., F. Willer, R. Samtleben, and G. Boos. 1994. *Phytomedicine* 1:213–224.
70. Willer, F., H. Wagner, and E. Schecklies. 1991. *Deutsche Apotheker Zeitung* 131:1217–1221.
71. Hryb, D. J., M. S. Khan, N. A. Romas, and W. Rosner. 1995. *Planta Medica* 61:31–32.
72. *Bundesanzeiger* (Cologne, Germany): January 5, 1989; March 6, 1990.
73. *Bundesanzeiger* (Cologne, Germany): April 23, 1987.
74. Marandola, P., H. Jallous, E. Bombardelli, and P. Morazzoni. 1997. *Fitoterapia* 68:195–204.
75. Schulz, V., R. Hänsel, and V. E. Tyler. 1998. *Rational phytotherapy,* 3rd ed., 232–233. Berlin: Springer–Verlag.

chapter five

Respiratory tract problems

Bronchial asthma

Bronchial asthma is a condition characterized by difficulty in breathing. It occurs when extrinsic factors (allergens) or intrinsic factors (nonimmunological conditions) cause various mediators, including histamine and the leukotrienes, to be released from mast cells and circulating basophils. This results in relatively rapid contraction of the smooth muscle that surrounds the airways, accompanied by a slower secretion of thick, tenacious mucus and edema of the respiratory mucosa. Characterized by wheezing, coughing, shortness of breath, and tightness in the chest, bronchial asthma is reversible. It is a complex disease in which airway obstruction is responsible for the clinical manifestations; however, bronchial hyperresponsiveness and underlying inflammation are also characteristic features. Treatment involves the use of bronchial dilators (the most common is theophylline) as well as various adrenergic amines and anticholinergic agents that reverse the acute attack (bronchoconstriction) or prevent bronchospasm from occurring. In addition, anti-inflammatory agents such as aerosol corticosteroids and cromolyn sodium modify the airway inflammation and reduce bronchial hyperresponsiveness.[1]

Patients with chronic bronchitis and emphysema, which are other forms of chronic obstructive pulmonary disease (COPD), experience some symptoms similar to those associated with asthma. In these conditions, however, the symptoms are usually continuous and not episodic as in asthma and should be treated only under the care of a physician.

Theophylline occurs with the related xanthine derivatives caffeine and theobromine in several different plant products, including coffee, tea, cocoa, and cola. The highest concentration occurs in tea, *Camellia sinensis* (L.) O. Kuntze, but even there it seldom exceeds 0.0004 percent.[2] Consumption of approximately 55 lb of tea would be required to equal one 100-mg tablet of theophylline. When one considers that six to eight times this quantity would be administered daily to relieve an acute attack of bronchial asthma, the impossibility of consuming tea as a source of the necessary theophylline becomes obvious.

Plants containing the solanaceous alkaloids atropine, hyoscyamine, and scopolamine reduce bronchospasm by their anticholinergic action.

Formerly, the leaves of Jimson weed (*Datura stramonium* L. of the family Solanaceae) were burned and the smoke inhaled to alleviate the condition. Such use has been largely discontinued, not only because of high potential toxicity but also because the alkaloids reduce bronchial secretion and ciliary activity of the bronchial epithelium, thus diminishing the expectorant action needed to clear the respiratory passages.[3]

The only common plant remedy useful for the treatment of bronchial asthma is the adrenergic herb ephedra.

Herbal remedies for bronchial asthma

Ephedra

Often referred to by its Chinese name, *ma huang,* ephedra (the green stems of various *Ephedra* species, particularly *E. sinica* Stapf, *E. equisetina* Bunge, and others of the family Ephedraceae) has been used in China for the treatment of bronchial asthma and related conditions for more than five thousand years. Another species, *E. gerardiana* Wall., has been similarly employed in India. Ephedra was the first Chinese herbal remedy to yield an active constituent—in this case, ephedrine—widely used in Western medicine.[4]

Ephedrine was first isolated from the herb by a Japanese chemist, N. Nagai, in 1887. Nearly forty years elapsed before K. K. Chen and his mentor, C. F. Schmidt, of the Peking Union Medical College began to publish (1924) a series of studies on the pharmacological properties of the alkaloid. American physicians were quick to appreciate the adrenergic properties of ephedrine, and it became widely used as a nasal decongestant, a central nervous system stimulant, and a treatment for bronchial asthma.[5] In addition to ephedrine, several other alkaloids, such as pseudoephedrine, norephedrine, and norpseudoephedrine, etc., are contained in various species of *Ephedra.* These possess physiological properties similar to those of ephedrine.[6]

The approximately forty different species of *Ephedra* are grouped by Hegnauer into five geographic types based primarily on variations in their alkaloid content.[7] All North and Central American types appear to be devoid of alkaloids; thus, any activity attributed to these species must derive from compounds other than ephedrine or its derivatives. This is of particular interest in the case of *E. nevadensis* S. Wats., the ingredient in Mormon tea. As an American species of *Ephedra,* it is alkaloid free and of no value in the treatment of bronchial asthma. It should be emphasized that even specialists often find the different species of *Ephedra* difficult to classify.

When a β_2-agonist binds to the receptor on the cell membrane, a conformational change in the receptor results in a signal transduction involving the participation of G proteins. Activation of adenylate cyclase occurs, leading to increased intracellular levels of cyclic AMP and a reduction in cytoplasmic Ca^{++} concentration, resulting in smooth-muscle relaxation. Ciliary activity and liquefaction of tenacious mucus also increase, resulting in a mild expectorant action.[8]

Ephedra is recommended for treating only mild forms of seasonal or chronic asthma. Due to its particular chemistry and more lipophilic properties compared to norepinephrine, ephedrine is effective when administered orally; its peak effect occurs in one hour after administration and it has a half-life of six hours. Its duration of action is more prolonged than for norepinephrine, which is not effective upon oral administration, because ephedrine is resistant to metabolism by both monoamine oxidase (MAO) and catechol-O-methyltransferase (COMT).

The herb is often administered in the form of a tea prepared by steeping 1 heaping teaspoonful (2 g) in ½ pint of boiling water for ten minutes.[9] Prepared from plant material of good quality, this would represent 15–30 mg of ephedrine, which approximates the usual dose of the alkaloid.

Ephedra has multiple, serious adverse effects (particularly in large doses), principally because ephedrine is a nonselective adrenergic agonist. Figure 5.1 shows the effects of ephedra on adrenergic receptors in various target tissues of the body and how they translate into side effects. Central nervous system stimulation can be a problem resulting in nervousness, insomnia, hyperactivity, and irritability. Its effects on the cardiovascular system include tachycardia, premature systoles, and vasoconstriction, causing an increase in both systolic and diastolic blood pressure. Its action in the gastrointestinal tract to decrease tone, motility, and secretory activity may result in nausea and vomiting. These side effects render the indiscriminate use of ephedra highly inadvisable, particularly by persons suffering from heart conditions, hypertension, diabetes, or thyroid disease. It would also be contraindicated in patients with prostatic hyperplasia (BPH) because of the problem of urinary retention.[10]

Various products containing ephedra or ephedrine have been marketed with numerous unsubstantiated and potentially dangerous therapeutic claims. These include weight loss, appetite control, and as an alternative to illegal street drugs such as "Ecstasy" (MDMA or 3,4-methylenedioxymethamphetamine). These claims have led to a plethora of cases of abuse with the serious consequence of more than six hundred reports in the last few years to the Food and Drug Administration (FDA) of ephedrine-related adverse reactions, along with twenty-two deaths attributed to the drug.[11] In an attempt to control this abuse, the FDA has proposed regulations that will control the amount of ephedrine alkaloids in a product and will not allow combination products containing caffeine or caffeine-containing herbs (also central nervous system stimulants).[12]

Because of its chemical structure, ephedrine can serve as a precursor for the illegal synthesis of methamphetamine or "speed," a common drug of abuse. Several states have recently passed laws regulating the sale

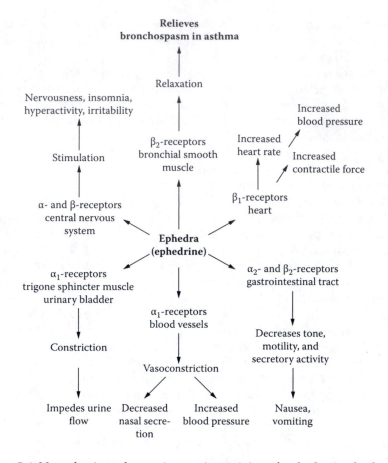

Figure 5.1 Nonselective adrenergic agonist activity of ephedra in the human body.

of the alkaloid or products containing it. This concern overlooks the fact that today most ephedrine is produced by a chemical synthesis involving the reductive condensation of l-1-phenyl-1-acetylcarbinol with methylamine. This yields the desired isomer l-ephedrine, which is identical in all respects to that contained in ephedra. In view of the difficulties involved in extracting and purifying the relatively small concentrations of ephedrine from the ephedra herb and the fact that the plant serves only as a minor source of the alkaloid, anyway, restricting the availability of the herb, although well intended, seems an excessive measure.[13,14]

Bitter or sour orange

Citrus aurantium L., family Rutaceae (also called Seville and Neroli orange), has been vigorously promoted as a substitute for ephedra in weight-loss products, following restriction of products of the latter herb by the FDA due to their clinical association with strokes, heart attack, hypertension, and psychiatric problems.[15] Extracts are prepared from the immature fruit of *C. aurantium*, termed *zhi shi* in Chinese. The extracts contain five adrenergic amines; chief among them is synephrine, accompanied by N-methyltyramine, hordenine, octopamine, and tryamine. The extracts are usually standardized to 4 or 6 percent synephrine.

In rats, reported oral administration of 2.5–20 mg/kg of two *C. aurantium* fruit extracts standardized to 4 and 6 percent synephrine, respectively, significantly reduced food intake and body weight gain. However, mortality (not observed in controls) was present in all *C. aurantium* groups. Although arterial blood pressure was not modified, ECG alterations (ventricular arrythmias with enlargement of QRS complex) were evident in animals treated with both extracts.[16]

The only published trial of a *C. auranitum*-containing weight-loss product utilized a combination of 975 mg *C. aurantium* extract (6 percent synephrine), 528 mg caffeine, and 900 mg of a St. John's wort extract (0.3 percent hypericins) daily for six weeks received by nine subjects. Seven received a placebo and four no treatment other than dietary counseling and an exercise program. The product was found not to be superior to placebo for weight loss.[17] There is also no evidence that synephrine and its accompanying sympathomimetic amines would have any lipolytic effect on human adipocytes at the levels found in weight-loss products.

Concern expressed over the potential for *C. aurantium* dietary supplements to affect drug metabolism[18] was judged groundless by the American Herbal Products Association. This organization pointed out that although bitter orange *juice* is claimed to be a potent inhibitor of cytochrome P450 3A4 isoenzyme and would therefore be expected to increase the blood level of many drugs, there is no evidence to support such effects from bitter orange *extracts* made from dried fruit or peel.[19]

Thus far, there have not been any reports of adverse effects associated with consumption of bitter orange dietary supplements.

Colds and flu

Acute viral infections of the upper respiratory tract produce a mixture of symptoms variously called the "common cold," acute rhinitis, or catarrh. Symptoms of this highly contagious condition include nasal congestion and discharge accompanied by sneezing, irritation, or a "tickling" sensation in the dry or sore throat that gives rise to cough, laryngitis, bronchial congestion, headache, and fever. If the infection is particularly severe and results in significant malaise—including joint and muscle pain and, possibly, gastrointestinal disturbances—the condition is called influenza or "flu." Both conditions are self-limiting (five to ten days) but may become complicated by secondary bacterial infections.[20] Further, some symptoms, such as cough, may persist for several weeks.

Treatment of the common cold and flu is largely symptomatic; curative remedies do not exist. In addition to ephedra, which may serve as a useful decongestant, the most effective herbal remedies are those used to treat coughs. These fall into two categories: antitussives (cough suppressants) and expectorants. The two are closely related, and there is some overlap of herbal products used to treat the condition.

Demulcent antitussives

Antitussives act either centrally on the medullary cough center of the brain or peripherally at the site of irritation. Although some of the best centrally active antitussives (e.g., codeine) are plant products, they are subject to abuse and are not available for self-selection. Consequently, they are not discussed here.

Certain volatile oils obtained from herbs are incorporated into a sugar base and marketed in the form of lozenges to suppress coughs. Some of the more popular oils used for this purpose include anise, eucalyptus, fennel, peppermint, and thyme. Cough drops flavored with these oils apparently function by stimulating the formation and secretion of saliva, which produces more frequent swallowing and thereby tends to suppress the cough reflex.[21] However, the real therapeutic utility of volatile-oil-containing herbs in treating conditions associated with colds and flu is their expectorant action. The effective herbal expectorants will be considered following the antitussives.

Two natural herbal products, camphor and menthol, are used topically as antitussives. Ointments containing these drugs are rubbed on the throat and chest, and they may also be used in steam inhalers. The aromatic vapor that is inhaled from these applications has a local anesthetic action on the lungs and throat, which suppresses the cough reflex. Menthol is used in cough drops for this same purpose.

The antitussive effect of many herbs results from the content of mucilage, which exerts a demulcent or protective action. Mucilages are hydrophilic colloids that, in the presence of water, tend to form viscous solutions—or tacky gels. When consumed, usually in the form of a tea, they form a protective layer over the mucous membrane of the pharynx, larynx, and trachea, thereby preventing mechanical irritation of the receptors there and preventing the cough reflex.

Because the mucilage is not absorbed and its action is essentially a mechanical one, it does not produce untoward side effects. However, some mucilage-containing herbs possess additional constituents that are toxic. This is the case with one of the long-used herbal antitussives, coltsfoot (the leaves of *Tussilago farfara* L. of the family Asteraceae). Although coltsfoot has useful cough-protective properties, its use cannot be recommended because it also contains toxic pyrrolizidine alkaloids (PAs).[22]

The following mucilage-containing antitussive herbs may be employed more or less interchangeably and are listed in alphabetical order.

Iceland moss

Not a higher plant, but rather a lichen—that is, an alga and a fungus growing in symbiotic association—Iceland moss is obtained from *Cetraria islandica* (L.) Ach. of the family Parmeliaceae. Commercial supplies of this foliaceous lichen are obtained primarily from Scandinavia and central Europe. It contains about 50 percent of a mixture of mucilaginous polysaccharides, principally lichenin and isolichenin. Iceland moss is consumed in the form of a decoction prepared from 1–2 heaping teaspoonfuls of herb and 150 mL of water. Drink one cup three times a day.[23] Total daily dose is 4–6 g of plant material. The German Commission E has found Iceland moss effective for the treatment of irritations of the mouth and throat and associated dry cough.[24]

Finnish scientists warn against utilizing Iceland moss in large quantities over an extended period of time. It has long been used as an emergency food in that country; however, in recent years, the lead content of the lichen has increased to the point (30 mg per kilogram of dry weight) at which this practice can no longer be considered safe.[25] Although the relatively small amounts used occasionally for the treatment of cough probably pose negligible risk, it is nevertheless a concern of which consumers should be aware.

Marshmallow root

This herb consists of the dried root, deprived of the brown outer corky layer, of *Althaea officinalis* L. (family Malvaceae). It contains 5–10 percent of mucilage and is consumed in the form of a tea, 1–2 teaspoonfuls in 150 mL of water (daily dose: 6 g), for its antitussive effect.[21] In Europe, the leaves of the plant, as well as the leaves and flowers of the common mallow, *Malva sylvestris* L., and related species and subspecies, are all employed similarly. Commission E has declared them all to be effective demulcents.[24]

Mullein flowers

The flowers of several species of mullein—*Verbascum thapsus* L., *V. densiflorum* Bertol., and *V. phlomoides* L. (family Scrophulariaceae)—all contain about 3 percent of a mucilage useful in the treatment of throat irritations and cough. The flowers (3–4 teaspoonfuls) are used to prepare 150 mL of tea, which may be drunk several times daily.[21] Approved by Commission E for the treatment of respiratory catarrh, the herb also has some expectorant activity.[26]

Plantain leaf

Fresh or dried leaves of the English plantain, *Plantago lanceolata* L. (family Plantaginaceae), have a worldwide reputation as a soothing cough suppressant. This action is attributed primarily to the approximately 6 percent of mucilage found in the plant material; tannins and bitter principles may contribute as well. The herb is also employed for inflammatory conditions of the oral cavity as well as to treat various skin inflammations.

Plantain's effectiveness in these latter conditions is due in part to its mucilage content; in addition, two iridoid glycosides, aucubin and catapol, almost certainly play a role, at least under certain conditions. When the sap is expressed from the fresh leaves, the glycosides are hydrolyzed, and the residual aglycones exert a strong antibacterial effect. This accounts for the folkloric use of the fresh crushed leaves as an anti-inflammatory and wound-healing agent. Of course, the antibacterial products are not present in the infusions customarily used to relieve coughs because boiling water inactivates the hydrolytic enzyme.[27]

German Commission E has found plantain safe and effective as a soothing demulcent, astringent, and antibacterial.[28] It is customarily administered as a tea prepared from 3–4 teaspoonfuls of the herb and 150 mL of boiling water.

Slippery elm

Indians and early settlers of North America valued the inner bark of the slippery elm, *Ulmus rubra* Muhl. (family Ulmaceae), as a poultice and soothing drink. The bark of this large tree, native to the eastern and central United States, contains large quantities of a viscid mucilage that acts as an effective demulcent and antitussive.[29] Although the herb may be consumed in the form of a tea, a number of throat lozenges containing it are commercially available. These are the preferred dosage form for the treatment of cough and minor throat irritations because they provide a sustained release of the mucilage to the pharynx.

This native American herb has not seen widespread usage elsewhere, so European authorities have not commented on its safety and efficacy. Some measure of its utility may be gathered from the fact that it was listed in the official compendia (USP and NF) from 1820 to 1960.[30] The FDA has declared it to be a safe and effective oral demulcent.

Expectorants

Prolonged irritation of the bronchioles results in an increase in the mucoprotein and acidic mucopolysaccharide content of their secretions and a concomitant increase in the viscosity of the mucus and other fluids. This and several related factors reduce the ability of ciliary movement and coughing to move the thickened secretions toward the pharynx.

Symptomatic therapy with expectorants has the objective of reducing the viscosity of these secretions so that the loosened material may be eliminated from the system, eventually, by expectoration.

The action of the so-called nauseant–expectorant herbs that contain alkaloids results primarily from their action on the gastric mucosa. This provokes a reflex stimulation of the vomiting center in the brain via the vagus nerve, which leads to an increase in secretion of the bronchial glands. Volatile-oil type expectorant herbs, on the other hand, exert a direct stimulatory effect on the bronchial glands by means of local irritation. Saponin-containing expectorant herbs function by reducing the surface tension of the secretions, facilitating their separation from the mucous membranes.[31] Of course, some expectorant herbs combine two or more of these effects. For example, the saponin-containing senega root also possesses nauseant–expectorant properties.

Use of expectorants is based primarily on tradition. Subjectively, they appear to be effective for the treatment of irritative, nonproductive coughs associated with a small amount of secretion. Substantial proof of their therapeutic utility is lacking. Nevertheless, they form a significant group of herbal remedies, and some appear to be of value. In the following discussion, they are classified on the basis of their mode of action: (1) nauseant–expectorants, (2) local irritants, and (3) surface-tension modifiers. The classification is imprecise because the function of many of the herbs is incompletely understood and some play multiple roles.

Nauseant–expectorants

The two most effective nauseant–expectorant herbs cannot be used extemporaneously because of their potential toxicity and the need to administer their active constituents in precise doses.

Ipecac
The first of these is ipecac, which consists of the rhizome and roots of *Cephaelis ipecacuanha* (Brot.) A. Rich. or *C. acuminata* Karst. of the family Rubiaceae.[32] Ipecac syrup (USP) is widely used as an emetic in the treatment of certain poisonings, but it is prepared from powdered ipecac (USP), which is standardized to contain from 1.9 to 2.1 percent of the active ether-soluble ipecac alkaloids—primarily emetine, caphaeline, and psychotrine. A number of commercial expectorant mixtures also contain precise amounts of standardized ipecac, and these are the preferred dosage form of this useful expectorant. Follow the directions on the label.

Lobelia
A second effective herbal nauseant–expectorant, which cannot be used safely because standardized preparations do not exist, is lobelia. Commonly

called Indian tobacco, it consists of the leaves and tops of *Lobelia inflata* L., family Campanulaceae. This native American plant was at one time widely used in the United States by so-called lobelia doctors who practiced a system of medicine developed by Samuel Thomson in the early nineteenth century. As a result of its contained alkaloids (principally lobeline), lobelia is an effective nauseant–expectorant, but the ratio of risk to benefit is very high. Its use as a crude herbal product is not recommended.[33]

Local irritants

Two effective volatile-oil-containing expectorant herbs, anise and fennel, have been discussed previously in Chapter 3, "Digestive System Problems." See that chapter for details.

Horehound

Possibly the most effective and pleasant-tasting plant drug in this category is horehound. Consisting of the leaves and flowering tops of *Marrubium vulgare* L. (family Lamiaceae), horehound has been used as a cough remedy for some four hundred years. It also has choleretic properties, so it serves to facilitate digestion as well. The activity of the herb is attributed not only to its content of volatile oil (0.06 percent) but also especially to a bitter diterpenoid lactone, marrubiin (or its possible precursor in the plant, premarrubiin). This compound exerts a direct stimulatory effect on the secretions of the bronchial mucosa.[34]

In 1989, the FDA banned horehound from over-the-counter cough remedies because it had not received sufficient evidence supporting its efficacy. It is still on the GRAS (generally recognized as safe) list, however, and in 1990 the German Commission E approved horehound for the treatment of bronchial catarrh as well as dyspepsia and loss of appetite.[21]

Horehound may be consumed as a tea prepared from 2 heaping teaspoonfuls of the cut herb steeped in ¼ L of boiling water. Three to five cups may be consumed daily; untoward side effects have not been reported.[35] The herb is also available in the form of hard horehound candy that is widely used as a cough lozenge.

Thyme

Another useful irritant–expectorant herb is thyme. The leaves and tops of two different species of *Thymus* (family Lamiaceae) are now used more or less interchangeably. These are *T. vulgaris* L. (common or garden thyme) and *T. zygis* L. (Spanish thyme). However, it must be noted that common thyme contains a greater quantity of volatile oil (0.4–3.4 percent) than Spanish thyme (0.7–1.38 percent), so equivalent weights are not equivalent therapeutically. The principal constituents of the oil are various phenols, especially thymol (30.7–70.9 percent) and carvacrol (2.5–14.6 percent). Flavonoids, tannins, and triterpenes are also present.[36]

The volatile oil not only has expectorant and antiseptic properties but also functions to relieve bronchospasm.[37] This spasmolytic effect is enhanced by the flavonoids in the plant.[38] Commission E has found thyme to be effective for the treatment of the symptoms of bronchitis, pertussis, and catarrh.[39] It is normally consumed as a tea prepared from 1–2 g of the herb per cup of water. One cup of the moderately warm tea is drunk up to three times daily. It may be sweetened with honey, which also acts as a demulcent, thereby increasing the tea's effectiveness.

Eucalyptus leaf

Although used relatively little in comparison to the volatile oil obtained from them, the leaves of *Eucalyptus globulus* Labill. (family Myrtaceae) and related species do possess a useful expectorant activity. To be effective medicinally, the leaf oil must contain 70–85 percent cineole (eucalyptol). This is an important criterion because although many species and chemical races of the genus yield 3–6 percent volatile oil, some of them do not contain sufficient cineole to provide the necessary expectorant and antiseptic activity.[40]

A tea prepared from 1 teaspoonful (1.5–2 g) of eucalyptus leaves in about 150 mL of hot water and drunk freshly prepared three times daily serves as a useful cough remedy. The volatile oil, which has official status in the NF, is commonly incorporated in a variety of nasal inhalers and sprays, balms and ointments (rubs) for external application, and mouthwashes.

Surface-tension modifiers

Of the small number of effective saponin-containing expectorant herbs, two are not commonly used in the United States. The leaves of ivy, *Hedera helix* L. (family Araliaceae), cannot be employed as a tea, and the concentrated extracts available in Europe are not articles of commerce in the United States. The flowers and root of primula, *Primula veris* L. or *P. elatior* (L.) Hill of the family Primulaceae, are very popular in Europe but not readily available in the United States.[41]

Licorice

Also known as glycyrrhiza, licorice is widely used in the United States and has very useful expectorant or antitussive properties. It has already been discussed as a treatment for stomach ulcers. Although licorice does contain saponins, its mode of action in the treatment of upper respiratory congestion and coughs requires considerable clarification.[41]

Senega snakeroot

The only significant herb remaining in this category is senega snakeroot. Variously known as senega or seneca root, this herb consists of the dried root of *Polygala senega* L. (family Polygalaceae). The plant is native to the eastern woodlands of North America. It was used by the Seneca Indians to treat rattlesnake bite—hence, the name.

Senega snakeroot contains 5–10 percent of a mixture of triterpenoid saponins, which are the active expectorant principles. The major components of the mixture are senegin, also known as polygalin, and polygalic acid. Although these probably function directly to reduce the viscosity of thickened bronchial secretions, their primary mechanism of action appears to be that of a nauseant–expectorant. Irritation of the gastric mucosa leads, by reflex stimulation, to an increase in bronchial mucous gland secretion.[42]

The herb is administered in the form of a decoction prepared from 0.5 g (about 1/5 teaspoonful) and 1 cup of water. Total daily dose should not exceed 3 g because of the tendency of large doses to upset the stomach and to produce nausea and diarrhea. Commission E has approved senega snakeroot as an expectorant for the treatment of upper respiratory catarrh.[43] The drug was official in the NF until 1960.

Sore throat

Often, but not necessarily, associated with colds and flu, sore throat may be a symptom of many illnesses. These range from acute simple (catarrhal) pharyngitis, usually caused by bacterial or viral infections of the upper respiratory tract, to severe streptococcal infections. It also accompanies certain acute specific infections, such as measles and whooping cough. The kind of dry sore throat that attends colds and flu is usually self-limiting; treatment is symptomatic with emphasis on increasing the patient's comfort. Gargling with warm infusions or decoctions of various herbs is often recommended. The antiseptic and astringent botanicals commonly used as palliatives are essentially the same as those employed for lesions and infections of the oral mucosa. They are discussed together in Chapter 10.

References

1. Torphy, T. J., and D. W. P. Hay. 1994. Drugs used in asthma. In *Modern pharmacology*, 4th ed., ed. C. R. Craig and R. E. Stitzel, 509–521. Boston: Little, Brown.
2. List, P. H., and L. Hörhammer, L., eds. 1973. *Hagers Handbuch der Pharmazeutischen Praxis*, 4th ed., vol. 3, 639. Berlin: Springer–Verlag.
3. Haas, H. 1991. *Arzneipflanzenkunde*, 64. Mannheim, Germany: B. I. Wissenschaftsverlag.

4. Osol, A., and G. E. Farrar, Jr. 1947. *The dispensatory of the United States of America*, 24th ed., 403–407. Philadelphia, PA: J. B. Lippincott.
5. Kreig, M. B. 1964. *Green medicine*, 415–416. Chicago: Rand McNally.
6. Steinegger, E., and R. Hänsel. 1988. *Lehrbuch der Pharmakognosie und Phytopharmazie*, 450–453. Berlin: Springer–Verlag.
7. Hegnauer, R. 1962. *Chemotaxonomie der Pflanzen*, vol. 1, 460–462. Basel: Birkhäuser Verlag.
8. Lee, T. J.-F., and R. E. Stitzel. 1994. Adrenomimetic drugs. In *Modern pharmacology*, 4th ed., ed. C. R. Craig and R. E. Stitzel, 115–128. Boston: Little, Brown.
9. Pahlow, M. 1985. *Das Grosse Buch der Heilpflanzen*, 387–388. Munich: Gräfe und Unzer.
10. Williams, D. M., and T. H. Seif. 1996. Asthma. In *Handbook of nonprescription drugs*, 11th ed., ed. T. R. Covington, 157–177. Washington, D.C.: American Pharmaceutical Association.
11. Blumenthal, M., and A. Dickinson. 1996. *HerbalGram* 38:28–31.
12. Blumenthal, M. 1997. *HerbalGram* 40:26–27.
13. *Health Foods Business* 37 (6): 8 (1991).
14. *Health Foods Business* 37 (8): 12 (1991).
15. Haller, C. A., and N. L. Benowitz. 2000. *New England Journal of Medicine* 343:1833–1838.
16. Calapai, G., F. Firenzuoli, A. Saitta, F. Squadrito, M. R. Arlotta, G. Constantino, and G. Inferrera. 1999. *Fitoterapia* 70:586–592.
17. Colker, C. M., D. S. Kalman, G. C. Torina, T. Perlis, and C. Street. 1999. *Current Therapeutic Research* 60:145–153.
18. Fugh-Berman, A., and A. Myers. 2004. *Experimental Biology and Medicine* 229.
19. American Herbal Products Association, Silver Spring, MD, September 2, 2004.
20. Tietze, K. J. 1996. Cold, cough, and allergy. In *Handbook of nonprescription drugs*, 11th ed., ed. T. R. Covington, 133–156. Washington, D.C.: American Pharmaceutical Association.
21. Hänsel, R. 1991. *Phytopharmaka*, 2nd ed., 99–104. Berlin: Springer–Verlag.
22. *Lawrence Review of Natural Products:* June 1996.
23. Bisset, N. G., ed. 1994. *Herbal drugs and phytopharmaceuticals*, English ed. (*Teedrogen*, M. Wichtl, ed.), 137–139. Boca Raton, FL: CRC Press.
24. *Bundesanzeiger* (Cologne, Germany): January 5, 1989.
25. Airaksinen, M. M., P. Peura, and S. Ontere. 1986. *Archives of Toxicology* 9 (suppl.): 406–409.
26. *Bundesanzeiger* (Cologne, Germany): February 1, 1990.
27. Bisset, N. G., ed. 1994. *Herbal drugs and phytopharmaceuticals*, English ed. (*Teedrogen*, M. Wichtl, ed.), 379–381. Boca Raton, FL: CRC Press.
28. *Bundesanzeiger* (Cologne, Germany): November 30, 1985.
29. *Lawrence Review of Natural Products:* March 1991.
30. Claus, E. P., and V. E. Tyler, Jr. 1965. *Pharmacognosy*, 5th ed., 85. Philadelphia, PA: Lea & Febiger.
31. Haas, H. 1991. *Arzneipflanzenkunde*, 65–72. Mannheim, Germany: B. I. Wissenschaftsverlag.
32. Robbers, J. E., M. K. Speedie, and V. E. Tyler. 1996. *Pharmacognosy and pharmacobiotechnology*, 160–161. Baltimore, MD: Williams & Wilkins.

33. Tyler, V. E. 1987. *The new honest herbal,* 150–151. Philadelphia, PA: George F. Stickley.
34. Tyler, V. E. 1987. *The new honest herbal,* 127–128. Philadelphia, PA: George F. Stickley.
35. Bisset, N. G., ed. 1994. *Herbal drugs and phytopharmaceuticals,* English ed. (*Teedrogen,* M. Wichtl, ed.), 317–318. Boca Raton, FL: CRC Press.
36. Bisset, N. G., ed. 1994. *Herbal drugs and phytopharmaceuticals,* English ed. (*Teedrogen,* M. Wichtl, ed.), 493–495. Boca Raton, FL: CRC Press.
37. Reiter, M., and W. Brandt. 1985. *Arzneimittelforschung* 35 (I): 408–414.
38. Van den Broucke, C. O., and J. A. Lemli. 1983. *Pharmaceutisch Weekblad,* scientific ed. 5: 9–14.
39. *Bundesanzeiger* (Cologne, Germany): December 5, 1984; March 6, 1990.
40. Tyler, V. E., L. R. Brady, and J. E. Robbers. 1988. *Pharmacognosy,* 9th ed., 133–135. Philadelphia, PA: Lea & Febiger.
41. Hänsel, R. 1991. *Phytopharmaka,* 2nd ed., 105–120. Berlin: Springer–Verlag.
42. Briggs, C. J. 1988. *Canadian Pharmaceutical Journal* 121:199–201.
43. *Bundesanzeiger* (Cologne, Germany): March 13, 1986; March 6, 1990.

chapter six

Cardiovascular system problems

Congestive heart failure

Congestive heart failure (CHF) is a relatively common clinical disorder in which the heart fails to provide an adequate blood flow to the peripheral tissues of the body. It is a serious condition experienced by some one-half million Americans; the five-year survival rate of such patients is less than 50 percent. CHF symptoms are associated with one or more of five key pathophysiologic features. These include blood pressure overload, volume overload, loss of heart muscle, decreased contractility, and disturbances in filling the heart. When such conditions result in reduced cardiac output, several compensatory mechanisms are activated that may sustain performance for a limited period, but without appropriate drug intervention, cardiac efficiency soon declines. The resulting symptoms include both ankle and pulmonary edema as well as ascites.[1]

The many etiologies and degrees of severity of heart failure require an individualized approach in treating each patient. Nevertheless, certain general principles of treatment apply to the management of various subsets of patients. Generally, heart failure is characterized by a slow, progressive decrease in cardiac function over many months or years, and many patients will initially have mild symptoms of heart failure that can be treated on an outpatient basis. Classifying patients according to their level of disability is useful in determining the course of treatment. The most-used classification system is the New York Heart Association (NYHA) Functional Classification System. This system divides patients into functional class: FC-I (patients with cardiac disease but without limitations of physical activity), FC-II (patients in whom physical activity results in fatigue, palpitations, dyspnea, or angina), FC-III (patients who have marked limitation of physical activity), and FC-IV (patients unable to carry on physical activity without discomfort).

The goals of therapy in heart failure are to reduce symptoms and hospitalizations, slow the progression of the disease process, and prolong survival. Reduction of the work load of the failing heart can be accomplished by physical and emotional rest as well as correction of obesity. In addition, one of the major compensatory responses to heart failure is sodium and water retention; therefore, restriction of dietary sodium is also an important nonpharmacologic approach to the management of heart failure. Drug therapy is usually initiated in patients with NYHA FC-II. In many cases, a mild diuretic is sufficient to restore the patient to an asymptomatic state. This therapy may be followed by a vasodilator such as an angiotensin-converting enzyme (ACE) inhibitor or the combination of hydralazine/isosorbide dinitrate. The cardioactive glycosides are the last group of drugs used as a mainstay in therapy, particularly in patients more seriously involved with the disease.

Herbs containing potent cardioactive glycosides

During this century, the drugs utilized most frequently by physicians for the treatment of CHF have been obtained from digitalis—the dried leaves of *Digitalis purpurea* L. or the dried leaves of *D. lanata* Erh. These members of the plant family Scrophulariaceae yield several potent cardiac glycosides, especially digitoxin, which is derived from both, and digoxin, which is prepared only from *D. lanata*. These two glycosides now account for all of the digitalis prescriptions normally dispensed in the United States. However, a standardized *D. purpurea* leaf preparation, powdered digitalis, is official in the USP and is still employed, along with similar products, in other countries.

Numerous other plants contain cardioactive glycosides with steroidal structures and physiological functions similar to those of digitalis. Some of these have been used from time to time in the treatment of CHF, but none presents any special advantage over digitalis. Therefore, only the names and botanical origins of some of the more common ones are listed here[2]:

- adonis—*Adonis vernalis* L., family Ranunculaceae;
- apocynum or Black Indian hemp—*Apocynum cannabinum* L. or *A. androsaemifolium* L., family Ranunculaceae;
- black hellebore—*Helleborus niger* L., family Ranunculaceae;
- convallaria or lily of the valley—*Convallaria majalis* L., family Convallariaceae;
- oleander—*Nerium oleander* L., family Apocynaceae;
- squill—*Urginea maritima* (L.) Bak., family Liliaceae; and
- strophanthus—*Strophanthus kombé* Oliv. or *S. hispidus* DC., family Apocynaceae.

CHF is a serious disturbance of multiple origins that requires prompt, accurate diagnosis and careful treatment. The cardiac glycosides are extremely potent drugs, the dosage of which must be carefully adjusted to the needs of the individual patient. In the case of phytomedicines containing them, this is possible only with the standardized powdered digitalis, a product not readily available in the United States. Because nonprofessional diagnosis and treatment of seriously involved CHF patients are not in the best interest of the patients, the names and sources of the herbal products containing cardioactive glycosides employed to treat it are listed here only for the record, and no further details concerning their use are provided.

Other herbs for treating CHF

Hawthorn

In Europe, an extract of the dried leaves and blossoms of hawthorn is widely used to treat the early stages of congestive heart failure and minor arrhythmias for which cardioactive glycosides are not yet indicated. In some cases, it is used as adjunct therapy in combination with cardioactive glycosides.

This herb consists of the leaves with flowers or fruits of *Crataegus laevigata* (Poir.) DC. or *C. monogyna* Jacq. of the family Rosaceae. The principal activity of these plant materials is attributed to their content of oligomeric procyanidins derived from catechin or epicatechin. Additional effects are provided by various flavonoids, including quercetin, rutin, hyperoside, vitexin, and vitexin rhamnoside.[3]

In chronic heart failure, hawthorn is reported not only to improve the pumping capacity of the heart but also to reduce patients' susceptibility to cardiac angina. In contrast to the cardioactive glycosides, which mainly act on the cardiac muscle, hawthorn acts both myocardially and peripherally to reduce vascular resistance. The vasodilating and positive inotropic effects of hawthorn extracts have been attributed to the oligomeric procyanidins and flavonoid constituents.[4,5]

However, accurate evaluation of the utility of hawthorn is difficult because most of the pharmacological and clinical studies of it have been conducted utilizing standardized extracts prepared by methods of which the details are proprietary information.[6] Still, it appears that the herb causes a direct dilation of the smooth muscles of the coronary vessels, thereby lowering their resistance and increasing blood flow. The tendency toward angina is thus reduced. It is, however, not useful for acute attacks because its effects develop quite slowly following continued consumption.

In a randomized controlled study testing the effects of daily doses of a commercial hawthorn extract, the test group of seventy-eight patients with NYHA FC-II congestive heart failure made significant gains in their stamina and endurance as measured by a stationary bicycle. Test patients also had lower blood pressure and heart rates while exercising and felt less fatigue and shortness of breath than controls.[7] Research studies presented at the Crataegus Symposium in Cologne, Germany, in 1994 reported pharmacological investigations demonstrating that hawthorn extract improves the perfusion of blood to all areas of the peripheral vascular system. Four independent clinical, placebo-controlled studies showed a connection between the length of treatment and therapeutic efficacy. Physical functioning capacity improved at daily doses of extract over a period of at least four weeks and, to an even greater degree, after eight weeks. It was also shown that the hawthorn extract was equivalent to the ACE inhibitor captopril in terms of influence on stress tolerance.[8]

The incidence of side effects in the clinical use of hawthorn is very low. The toxicity of the plant has been estimated at an LD_{50} of 4,000 mg/kg and can be considered quite safe if one compares this value with caffeine, which has an LD_{50} of 200 mg/kg.[9] A hydroalcoholic hawthorn extract standardized to 18.75 percent oligomeric procyanidins has been investigated in single- and repeat-dose toxicity studies. At one hundred times the recommended human dose, there was no target organ toxicity and a battery of standard mutagenic and clastogenic tests was negative.[10]

German Commission E has approved the use of hawthorn for cardiac insufficiency corresponding to NYHA FC-I and FC-II for patients with a feeling of pressure and tightness in the cardiac region, as well as for the aging heart not yet requiring cardiotonic glycosides. The minimal daily dosage is established on the basis of flavone (5 mg) calculated as hyperoside, total flavonoids determined as total phenols (10 mg), or oligomeric procyanidins calculated as epicatechin (5 mg).[11] Because standardized extracts providing these dosage levels are not available in the United States and the wisdom of self-treating any abnormal heart condition is highly questionable, the use of such a remedy—even one as devoid of side effects as this—cannot be recommended at this time. In the future, however, because of its demonstrated efficacy and low incidence of side effects, hawthorn has considerable potential as an important drug for patients with the milder symptoms of congestive heart failure, provided standardized preparations become available and they take the drug only after consultation with their physicians.

Arteriosclerosis

Hyperlipoproteinemia, commonly known simply as "high cholesterol," refers to the concentration of protein-bound cholesterol and triglycerides in the blood plasma; it is one of the major risk factors predisposing persons to arteriosclerosis. Three other major factors are diabetes mellitus, smoking, and hypertension. Arteriosclerosis tends to be a generalized condition involving all major arteries, to some degree, with critical involvement of only a few. It is characterized by a gradual narrowing and ultimate occlusion of the affected vessels, often accompanied by weakening of the arterial walls.

The type of arteriosclerosis characterized by discrete deposits of fatty substances (atheromatous plaques) in the arteries and by fibrosis and calcification of their inner layer is called atherosclerosis. Atherosclerotic involvement of the coronary arteries is known as coronary artery disease. It may lead eventually to a sufficient obstruction of the vessels to produce ischemic heart disease accompanied by angina pectoris and, subsequently, myocardial infarction (heart attack). Although any artery may be affected, other prime targets of atherosclerosis are the aorta and the cerebral arteries; the major consequences are aortic aneurysms and cerebral infarction (stroke), respectively.

The death rate related to atherosclerosis in the United States is among the highest in the world, with nearly 50 percent of all deaths attributed to this pathological condition.[12] Fortunately, it is now widely recognized that atherosclerosis is not only preventable but also amenable to treatment by using drugs that lower levels of blood serum cholesterol. In addition to numerous prescription drugs available for this purpose, several herbal medicines have demonstrated effectiveness in lowering serum cholesterol.

It is interesting that one of these herbs not only reduces cholesterol levels but also enhances blood fibrinolytic activity and inhibits platelet aggregation. There is some evidence to support the claim that it may even lower blood pressure. All of these actions would be beneficial in the prevention and treatment of arteriosclerosis and its consequences.

Herbal remedies for arteriosclerosis

Garlic

Consisting of the bulb of *Allium sativum* L., family Liliaceae, garlic has been consumed as a food and a medicine since the time of the Egyptian pharaohs and the earliest Chinese dynasties. It has been extensively investigated both scientifically and clinically. Well over one thousand papers on garlic and related alliums have been published in the last twenty years.[13]

As illustrated in Figure 6.1, the intact mesophyll storage cells of garlic contain an odorless, sulfur-containing amino acid derivative known as alliin [(+)-S-allyl-l-cysteine sulfoxide]. When the cells are crushed, it comes into contact with the enzyme alliinase located in neighboring vascular bundle sheath cells and is converted to allicin (diallyl thiosulfinate). Allicin is a potent antibiotic, but it is also highly odoriferous and unstable. Depending on the conditions (steam distillation, oil maceration, etc.), it can yield a number of other strong-smelling sulfur compounds, such as various diallyl sulfides, including mono-, di-, tri-, tetra-, penta-, and hexa-; various methyl allyl sulfides, including mono-, di-, tri-, tetra-, penta-, and hexa-; various dimethyl sulfides, including mono-, di-, tri-, tetra-, penta-, and hexa-; 2-vinyl-4*H*-1,3-dithiin, 3-vinyl-4*H*-1,2-dithiin, E-ajoene, and Z-ajoene.[14]

The ajoenes are apparently responsible for much of the antithrombotic properties of garlic. Aside from them and the known antibiotic activity of allicin, connections between specific chemical compounds yielded by garlic and its therapeutic properties are not necessarily well established.

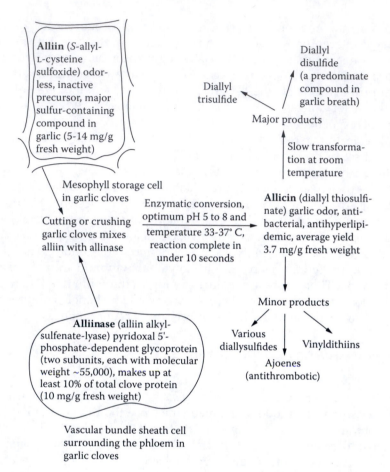

Figure 6.1 Transformations and origins of bioactive constituents of garlic.

However, an extensive list of tentative attributions has appeared.[15] Allicin is described there as possessing antiplatelet, antibiotic, and antihyperlipidemic activity. In consequence, most authorities now agree that the best measure of the total activity of garlic is its ability to produce allicin, which, in turn, results in the formation of other active principles. This ability is referred to as the allicin yield or potential of the garlic preparation.

Dutch investigators have evaluated the methodologies and results of eighteen controlled trials dealing with the beneficial effects of garlic on presumed cardiovascular risk indicators in humans: specifically, reduction of cholesterol, increased fibrinolytic activity, and inhibition of platelet aggregation. They concluded that the claims are valid for fresh garlic, but only at relatively high dosage levels. Most studies involved ingestion of 0.25–1 g of fresh garlic per kilogram of body weight per day. This is equivalent to a range of approximately five to twenty average-sized (4 g) cloves of garlic daily for a 175-lb person. Results with commercial garlic preparations were equivocal, as might be expected from dosage forms so diverse in their mode of preparation and constituents. Interestingly, fresh onions also yielded contradictory results, excepting an increased fibrin-olytic activity, which was consistently observed.[15]

In spite of their acceptance of the positive results, these investigators were highly critical of the methodology of most of the eighteen studies. However, under the strict guidelines applied by them, it would not be pos-sible to justify the validity of consuming any drug that lowered cholesterol, inhibited platelet aggregation, or increased fibrinolytic activity as a means of ameliorating arteriosclerosis and coronary artery disease. The impor-tant conclusion is that fresh garlic in large amounts does produce all of the effects believed by many authorities to be useful preventive measures.

Since 1975, there have been more than thirty-two human studies dem-onstrating the lipid-lowering effects of garlic. The majority of these studies were randomized, double blind, and placebo controlled; they lasted four to sixteen weeks and used hyperlipidemic patients. Fifteen studies uti-lized garlic powder tablets in daily doses ranging from 600 to 900 mg. The tablets were standardized to contain 1.3 percent alliin, which corresponds to an allicin release of 0.6 percent because 2 mol of alliin are required to produce 1 mol of allicin. Consequently, the 900-mg dose would have the potential of providing approximately 5.4 mg of allicin.

A summary of these studies showed an average serum cholesterol reduction range of 6–21 percent and an average serum triglyceride reduc-tion range of 8–27 percent.[16] These results also indicate that much smaller doses of fresh garlic than the five or more cloves daily suggested in the Dutch evaluation are effective in the treatment of hyperlipidemia. The German Commission E recommends 4 g (approximately one average-size clove) of fresh garlic as the daily dose for reducing elevated blood lipid levels. This amount is clearly in excess of the amount shown in clinical studies to have antihyperlipidemic effectiveness. Assuming that fresh garlic yields, on the average, 0.37 percent allicin, the 5.4-mg allicin poten-tial contained in the 900 mg of garlic powder ingested daily is equivalent to only 1.5 g of fresh garlic—less than one-half of an average-sized clove.

Discussing the effectiveness of commercial garlic preparations is like talking about the cause of cancer. They are so variable in their mode of prep-aration and resulting constituents that each would have to be addressed individually. However, there are a few basic principles. In preparing garlic powder, if the cloves are frozen, pulverized, and then vacuum-dried, the alliinase activity of fresh garlic is preserved. Because of expense, however, most powdered garlic is prepared by oven-drying large pieces of chopped cloves. Chopping causes a small, partial but immediate release of alliinase-generated thiosulfinates (e.g., allicin), which are lost upon drying. Alliinase and alliin are stable to temperatures of 60°C.[17]

The dried pieces of garlic can then be pulverized and alliin mixed with alliinase without fear of producing allicin because alliinase is inactive in the dried state. When the powder comes in contact with moisture (e.g., upon ingestion), alliinase is activated and rapidly converts alliin to the active allicin. But alliinase is inactivated by acids, so no conversion to allicin occurs in the stomach. Fresh garlic quickly releases allicin in the mouth during the chewing process, rather than in the stomach.[18]

Dried garlic preparations are most effective if the tablets or capsules are enteric coated so that they pass through the stomach and release their contents in the alkaline medium of the small intestine, where enzymatic conversion to allicin can readily occur. Once released, the allicin reacts rapidly with the amino acid cysteine derived from proteinaceous food consumed with the garlic. The S-allylmercaptocysteine thus formed effectively binds the odoriferous allicin, preventing it from reaching the bloodstream as such. When administered in this way, carefully dried garlic powder is probably relatively effective but produces little garlicky taste or odor.[19]

The activity of other garlic preparations is questionable, particularly those with an oil base. Allicin is unstable in oil, so much of the sulfide content and, consequently, much of the activity of such garlic preparations are lost. A Japanese preparation consisting of minced garlic aged in aqueous alcohol for eighteen to twenty months contained no detectable levels of allicin. In water, allicin is slowly converted to a number of volatile polysulfides that are also present in the steam-distilled oil. That oil, although high in total sulfides, contains neither allicin nor ajoene. It is obvious that the therapeutic value of various commercial garlic preparations, which appears to be directly related to the product's allicin yield, is highly dependent on the method of preparation, details of which are ordinarily unavailable.[20]

A 1992 study in Germany of eighteen of the approximately seventy garlic preparations commercially available in that country revealed that only five produced an allicin yield equivalent to 4 g of fresh garlic. That is the amount of fresh garlic, or its equivalent, established by German health authorities as the average daily dose required for therapeutic utility. The other thirteen preparations were so lacking in active principles as to be designated "expensive placebos." These findings are particularly significant in view of the fact that, in 1990, the garlic preparation market in the German Federal Republic amounted to approximately $160 million. It is estimated that 12 percent of all German citizens over the age of fourteen now consume garlic in prepared dosage forms.[21]

In 2001, Lawson and Wang evaluated all twenty-four known brands of enteric-coated garlic tablets for ability to release the claimed amount of allicin under simulated gastrointenstinal conditions. Nearly all brands employed effective coatings to protect the allicin-producing enzyme allii-nase from gastric acid inactivation. However, all brands but one showed low-dissolution allicin release when USP dissolution method 723 A was applied; 83 percent of the brands released less than 15 percent of their potential. The low allicin release was judged to be due to impaired allii-nase activity, mostly caused by excipients, and to slow tablet disintegration, which also impairs alliinase activity. The researchers recommend that garlic powder supplements be standardized on dissolution allicin release rather than on allicin potential.[22]

Since 1993, 44 percent of clinical trials of garlic have indicated a reduction in total cholesterol and a pronounced action to inhibit the ability of blood platelets to aggregate. Mixed results were obtained regarding garlic's effect on blood pressure and oxidative stress. A critical review published in 2006 assessed the significance of garlic and its constituents in cardiovascular disease. The authors conclude that although garlic appears to hold promise in reducing parameters associated with cardiovascular disease, "more in-depth and appropriate studies are required."[23] They particularly stress the importance of using standardized preparations and of establishing the bioavailability of active constituents. Well-designed randomized trials should be of sufficient duration to allow adequate assessment of morbidity, mortality, lipid, and thrombotic outcomes. Also important is to investigate whether garlic taken as a dietary supplement can either delay or prevent cardiovascular disease in a healthy population.

Consumption of moderate amounts of garlic does not pose a health risk for normal persons. However, based on garlic's antithrombotic activity and ability to inhibit platelet aggregation, there is a potential drug interaction when garlic is taken in combination with anticoagulant drugs and nonsteroidal anti-inflammatory drugs. The larger quantities thought by some to be required for therapeutic purposes (in excess of five cloves daily) can result in heartburn, flatulence, and related gastrointestinal problems. Allergies have also been reported, and those taking other anticoagulant drugs should consume garlic with caution.

In addition to its effectiveness in reducing some of the risk factors associated with arteriosclerosis and, specifically, coronary artery disease, garlic may have utility in the treatment of digestive ailments, bacterial and fungal infections, hypertension, and even cancer.[24,25] Additional studies on the constituents and therapeutic utility of this interesting plant are certainly warranted. At present, its use is approved by the German Commission E to support dietetic measures for the treatment of hyperlipoproteinemia and to prevent age-related changes in the blood vessels (arteriosclerosis).[26]

Green tea extract
Green tea extract has recently received considerable attention as a protective agent against cancer and cardiovascular disease. Use of green tea

extract for cancer protection purposes will be discussed in Chapter 11, "Performance and Immune Deficiencies." In the case of hyperlipoproteinemia, studies in rats have shown that ingestion of green tea extract significantly decreased plasma cholesterol and triglyceride concentrations and the ratio of low-density lipoprotein (LDL) cholesterol concentrations to high-density lipoprotein (HDL) cholesterol concentrations. A recent cross-sectional study in Japanese males surveyed on their living habits, including daily consumption of green tea, found that increased consumption of green tea (more than ten cups per day) was significantly associated with lower serum concentrations of LDL cholesterol and triglycerides and an increase in HDL cholesterol.[27]

Green tea contains antioxidant polyphenols including epicatechin, epicatechin gallate, epigallocatechin-3-gallate (EGCG), and proanthocyanidins. These antioxidants may be important in preventing the oxidation of LDL cholesterol, which plays a central role in the development of atherosclerosis.[28] Additional animal and human studies are required to establish firmly the efficacy of green tea extract in the prevention of atherosclerosis; however, because of the established safety of green tea, even in quantities of up to twenty cups per day, this herbal remedy holds considerable promise for cardiovascular disease prophylaxis.

Capsules containing dried green tea extract that has been standardized for antioxidant polyphenol content are available and should be taken with food and water. Those that are standardized to 97 percent polyphenols are said to be equivalent to four cups of green tea.

Red yeast

Red yeast is the rice fermentation product of a mixture of several species of *Monascus* fungi, principally *Monascus purpureus* Went (family Monascaceae). This product has been used for centuries in China as a food additive and medicinal agent; the first documented record dates from a.d. 800 during the T'ang Dynasty. The ancient Chinese pharmacopoeia characterizes red yeast as useful for the treatment of indigestion and diarrhea and for improving blood circulation and the health of the spleen and stomach. In China, it is still used in traditional medicine for these same purposes, and the fungus is important in the fermentation process to make rice wine and as a food coloring in preparing such delicacies as Peking duck. Considerable interest has been expressed in exploiting the potential of *M. purpureus* as a nitrate/nitrite substitute for the preservation of meats.[29]

The therapeutic activity of red yeast in hyperlipoproteinemia is due to monacolin K,[30] also designated lovastatin or mevinolin. It is representative of a group of functionalized hexahydronaphthalene β-hydroxy-δ-lactone compounds formed by certain fungi and collectively termed mevinic acids. These compounds are inhibitors of β-hydroxy-β-methylglutaryl CoA (HMG-CoA) reductase, the rate-limiting enzyme in endogenous

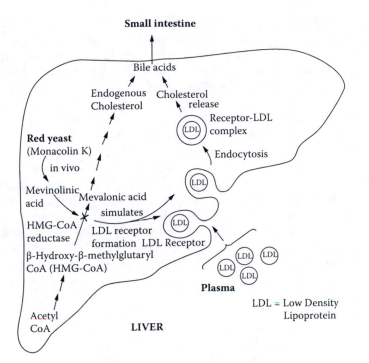

Figure 6.2 Mechanism of action of red yeast (monacolin K).

cholesterol biosynthesis (see Figure 6.2). Monacolin K is a prodrug, and in the body the lactone ring is hydrolyzed to mevinolinic acid, which resembles the chemical structure of the β-hydroxy-β-methylglutaryl portion of HMG-CoA. The mevinolinic acid binds to the HMG-CoA reductase rather than the natural substance HMG-CoA, blocking its conversion to mevalonic acid and the subsequent formation of cholesterol.

Normally, the cells of the body synthesize cholesterol de novo for use in cell-membrane structure and steroid-hormone synthesis. If additional cholesterol is required, the cell obtains it from circulating plasma LDL via receptor-mediated endocytosis. Figure 6.2 illustrates the de novo cholesterol synthesis in the liver, which requires cholesterol as a precursor for bile acid synthesis. The blocking of cholesterol biosynthesis by red yeast (monacolin K) stimulates hepatocytes to form a greater number of LDL receptors. This in turn promotes an increased influx of LDL cholesterol from the plasma to serve the precursor need in bile acid synthesis and a resulting decrease in plasma LDL cholesterol levels.[31]

The major active constituent of red yeast, monacolin K, is a well-recognized prescription drug under the U.S. adopted name (USAN) lovastatin. In healthy adults who maintained their usual diet, lovastatin produced mean reductions in serum total and LDL cholesterol of 23–27 percent and 35–45 percent, respectively, at dosages of 6.25–50 mg twice daily. The reductions were maximal within two weeks in most cases.[32] Because the active constituent in red yeast is lovastatin, it is not surprising that red yeast produces the same therapeutic profile as lovastatin and that the adverse-effect profile, precautions, and contraindications are the same as those of lovastatin.

Several clinical studies conducted in China and the United States support the efficacy of red yeast rice as a cholesterol-lowering agent. A double-blind, placebo-controlled, randomized clinical trial evaluated the safety and efficacy of a red yeast rice extract (Cholestin™, Pharmanex Inc.) in eighty-three otherwise healthy Americans, ages thirty-four to seventy-eight, with untreated hyperlipidemia. After eight weeks, total serum cholesterol levels in the red yeast rice-treated group decreased by 18 percent ($p < 0.001$), while those in the placebo group were unchanged; the extract group also exhibited reductions in LDL cholesterol and triglycerides, but HDL cholesterol was not significantly changed.[33] A legal dispute between Pharmanex and Merck, the manufacturer of the pharmaceutical lovastatin, prompted Pharmanex to replace the red yeast rice component of cholestin with policosanol, a cholesterol-lowering substance first extracted from sugar cane wax.[34] The legal issues attached to the dispute have been reviewed in Dumoff.[35]

The most frequent adverse effects reported in clinical trials with red yeast were gastritis, abdominal discomfort, and elevated levels of liver enzymes. Lovastatin has produced muscle pain, tenderness, and weakness, and there is the possibility that these may occur with red yeast use. Patients with existing liver disease or at risk for liver disease should not take red yeast; because of the importance of cholesterol in fetal development, pregnant women should not take the drug. In addition, children under age eighteen are not likely to benefit from cholesterol-lowering therapy, and safety has not been established in this age group.

Red yeast is available in capsules that contain 600 mg of a yeast–rice mixture standardized to supply 2.4 mg of HMG-CoA reductase inhibitor. The recommended dose is two 600-mg capsules taken twice daily with meals (2,400 mg/day).[36]

Policosanol

In addition to Cuban policosanol (CPC)—a mixture of many cosanols, being long-chain aliphatic alcohols (twenty-two to thirty-eight carbon atoms long)—other policosanols derive from wheat germ, rice bran, beeswax, sorghum, and even flax waste stream.[37] Most of the research on policosanol was originally conducted on CPC on almost exclusively Latino subjects from Central and South America. A review in 2002 of more than twenty randomized controlled trials (RCTs) described CPC as very well tolerated, safe, natural, and "a fascinating new agent for the prevention and treatment of atherosclerotic disease."[38] However, a later study with

German patients using a generic non-CPC (gPC) wheat germ-derived policosanol—with a chemical profile quite different from that of CPC—found no changes in blood lipids after four weeks of treatment.[39]

Other RCTs—one in the United States using a rice-derived gPC with a quite different chemical profile from that of CPC[40] and another gPC from South Africa using a Chinese-produced sugar cane wax with a chemical profile similar to that of CPC[41]—similarly showed no significant changes in blood lipid parameters. Finally, Canadian researchers in two studies, the latter of which was a widely publicized study in the *Journal of the American Medical Association,* also found that none of the lipid parameters examined changed significantly in any of the tested subject groups.[42] Almada has noted[37]: "The evidence base supporting CPC has eroded dramatically since its 'global birth' at the turn of the 21st century. The ostensible promise and the eventual disappointment underscore the need for independent RCTs and for multiethnic interventions."

Peripheral vascular disease

Peripheral vascular disease (PVD) is a general term that includes any disease of the blood vessels outside the heart and thoracic aorta and also disease of the lymph vessels. Three specific types are of special concern because they may respond to treatment with herbal remedies. These are (1) cerebrovascular disease, (2) other peripheral arterial circulatory disturbances, and (3) venous disorders.[43]

Cerebrovascular disease

Cerebrovascular disease results from abnormalities of the vessels, such as atherosclerosis or arteritis, or from abnormalities of blood flow or of the blood itself. Changes in blood flow result not only from disease of the vessels but also from thrombotic or embolic processes. Such changes in the brain may produce a reduction in the blood flow—a condition designated *ischemia.*

The degree of ischemia and its duration vary greatly. Figure 6.3 illustrates in a simplified manner the sequence of events that can lead to the various types of cerebrovascular disease due to the consequences of ischemia. Diseases in this category represent the third leading cause of death (after heart disease and cancer) in the United States. The brain requires 20 percent of the total body oxygen consumption; because there is no reserve of oxygen in the brain, impairment of blood flow in the brain due to ischemia will result in a lack of oxygen (hypoxia). This damages brain cells, leading to a diverse range of pathological conditions. Depending on severity, location, and duration of the ischemia, these conditions can include stroke, transient ischemic attack (TIA), cerebral edema, vascular (multi-infarct) dementia, and milder unpleasant symptoms frequently experienced by

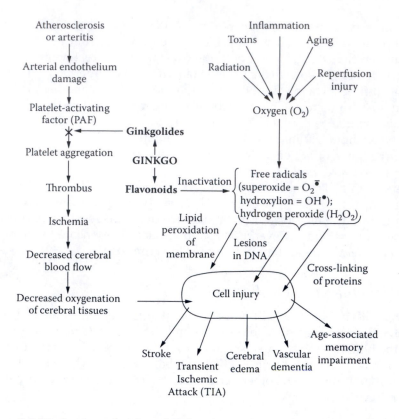
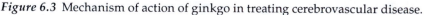

Figure 6.3 Mechanism of action of ginkgo in treating cerebrovascular disease.

elderly persons, including dizziness, depression, tinnitus, and short-term memory loss.[44]

In addition to brain cell damage due to hypoxia, cell injury can be a result of the formation of intermediate free radicals and other activated oxygen species generated from molecular oxygen during the course of normal physiologic processes such as inflammation and aging or by external insults to the body arising from radiation and toxins (see Figure 6.3). Free radicals have a single unpaired electron in an outer orbit and are extremely unstable and reactive. They readily attack double bonds in membrane polyunsaturated lipids, react with thymine in nuclear and mitochondrial DNA producing single-strand breaks, and promote sulfhydryl-mediated protein cross-linking, resulting in degradation or loss of enzymatic activity.[45]

Ginkgo
An herb that has shown considerable promise in alleviating many of these symptoms is ginkgo. A concentrated extract of the leaves of *Ginkgo biloba*

L., family Ginkgoaceae is currently enjoying enormous popularity in Europe as a treatment for peripheral vascular disease, particularly cerebral circulatory disturbances and certain other peripheral arterial circulatory disorders. Although the ginkgo tree has a very old history, having survived unchanged in China for some 200 million years, the herbal remedy prepared from its leaves is quite new, having been developed in the last twenty years. The leaves themselves cannot be used as an extemporaneous herbal remedy—for example, in the form of a tea—because they would provide an insufficient quantity of the active principles. As noted, *Ginkgo biloba* extract (GBE) is widely used in Europe; in 1995, the most frequently prescribed herbal monopreparation in Germany was GBE, which represented sales of 284 million U.S. dollars.[45] GBE is also available in Germany as an over-the-counter drug.

To supply the market with adequate supplies of leaves, plantations of ginkgo trees, severely pruned to shrub height to allow mechanical picking, have been established. A plantation in Sumter, South Carolina, comprises ten million ginkgos on 1,000 acres.[46] The leaves are picked green, dried, and shipped to Europe for processing. There, an acetone–water extract is prepared, dried, and adjusted to a potency of 24 percent flavonol glycosides and 6 percent terpene trilactones (ginkgolides and bilobalide).

Although complete analyses of GBE have not been conducted, the extract does contain a number of flavonol and biflavone glycosides, principally glycosides of quercetin and kaempferol; rutin is also present. The extract also contains 3 percent or more of a group of unique, closely related, bitter, twenty-carbon diterpene lactone derivatives known as ginkgolides A, B, C, J, and M. In addition, about 3 percent of a similar fifteen-carbon sesquiterpene designated bilobalide is present, together with 6-hydroxykynurenic acid, shikimic acid, protocatechuic acid, vanillic acid, and *p*-hydroxybenzoic acid.[47]

The therapeutic effects of GBE are attributed to a mixture of these constituents, rather than to a single chemical entity. Flavonoids of the rutin type reduce capillary fragility and increase the threshold of blood loss from the capillary vessels. This tends to prevent ultrastructural (ischemic) brain damage. They also function as free-radical scavengers and tend to inhibit the lipid peroxidation of cell membranes, to prevent oxidative modification of cellular proteins resulting in inactivation through cross-linking, and to prevent lesions in DNA that could lead to cell death or malignant transformation of cells.

Ginkgolides inhibit platelet-activating factor (PAF; see Figure 6.3). Produced by a variety of tissues, PAF not only induces aggregation of the blood platelets but also causes bronchoconstriction, cutaneous vasodilation, chemotaxis of phagocytes, hypotension, and the release from phagocytes of inflammatory compounds such as leukotrienes and prostaglandins as well as liposomal enzymes and superoxide. PAF appears to exert its actions by stimulating G protein-linked, cell-surface receptors. This causes activation of phospholipase A_2 with resultant formation of arachidonic acid, which is converted to prostaglandins, thromboxane A_2, and leukotrienes. In addition, the binding of PAF to its receptor on platelets unmasks cell-surface binding sites for fibrinogen that promote platelet aggregation and thrombus formation directly.[41] In short, it mimics many of the features seen in allergic response. All of these actions of PAF are blocked by the ginkgolides, particularly ginkgolide B. Possibly, the ginkgolides' most significant effect is an increase in blood fluidity, thereby improving circulation. Bilobalide acts in concert with the ginkgolides to improve the tolerance of brain tissue to hypoxia and to increase cerebral circulation.

Despite ginkgolides' acting as potent inhibitors of PAF, no clinical effects in humans have yet been observed. An anti-PAF effect, possibly intensified by inhibition of TXA2-dependent platelet aggregation conferred by ibuprofen, has been suggested as responsible for fatal intracerebral mass bleeding in a seventy-one-year-old man.[42] More recently, the importance of the PAF antagonistic effect of ginkgolides in hemorrhage associated with consumption of GBE has been questioned. It was noted that inhibition of aggregation of human platelets requires ginkgolide concentrations more than one hundred times greater then peak plasma levels measured after oral intake of a standardized proprietary extract (EGb 761) at recommended daily doses between 120 and 240 mg.[43] A systematic review of fifteen case reports describing a temporal association between bleeding and ginkgo consumption has appeared; only six reports clearly observed that bleeding did not recur after ginkgo use was terminated.[44]

Federal health authorities in Germany have declared GBE to be an effective treatment for cerebral circulatory disturbances resulting in reduced functional capacity and vigilance. Some of the symptoms resulting from such disturbances are vertigo, tinnitus, weakened memory, disturbances in concentration, and mood swings accompanied by anxiety. A primary target group is patients suffering from dementia syndromes, including primary degenerative dementia or vascular (multi-infarct) dementia.

Two recent clinical studies confirmed the efficacy of GBE in dementia of the Alzheimer type and vascular dementia. One investigation involved 216 outpatients in a prospective, randomized, double-blind, placebo-controlled, multicenter study over a twenty-four-week treatment period with a daily dose of either 240 mg GBE or placebo.[45] The second study utilized 309 patients over fifty-two weeks assigned randomly to treatment with GBE (120 mg/day) or placebo. Although modest, the changes induced by GBE were objectively measured by the Alzheimer's Disease Assessment Scale-Cognitive subscale and were of sufficient magnitude to be recognized by the caregivers.[46]

German Commission E has also found the extract useful for the treatment of certain other types of PVD—specifically, the peripheral arterial circulatory disturbance known as intermittent claudication. This condition, caused by sclerosis of the arteries of the leg, is characterized by a constant or cramping pain in the calf muscles brought on by walking a short distance. It is not uncommon among older persons and apparently responds well, at least in the initial stages, to treatment with GBE. A meta-analysis of five placebo-controlled clinical trials with GBE in which the effectiveness of treatment was measured by increases in pain-free walking distance showed significant improvement in patients after three to six months of treatment with 120 mg of GBE per day.[47]

A comprehensive review of the scientific literature that has examined the neurophysiological efficacy of GBE in healthy and cognitively intact adults appeared in 2005.[48] The majority of studies examining the acute effects of GBE in healthy adults found the extract efficacious in enhancing participants' neuropsychological functioning, particularly performance on tasks assessing attention, memory, and speed of processing. In two placebo-controlled, double-blind parallel studies, acute and chronic effects of a standardized GBE product on attention memory and executive function were examined in a well-defined population of healthy young university students (ages eighteen to twenty-six). The acute effects of a single dose of ginkgo significantly improved performance on the sustained-attention and pattern-recognition memory tests but did not significantly improve performance on the measures of working memory, planning, mental flexibility, and mood. Chronic treatment (six weeks) had no significant effect on mood or any of the cognitive tests.[49]

Side effects of GBE administration are neither numerous nor frequent. They may include gastrointestinal disturbances, headache, and allergic skin reactions. A case report of a hemorrhagic complication during GBE use in a patient also taking aspirin suggested that there may be an increase in bleeding tendencies if GBE is taken along with other drugs that inhibit blood platelet aggregation.[50]

The extract is sold in Europe as an approved drug and is available there in a variety of dosage forms, including tablets, liquids, and parenteral preparations for intravenous administration. It is not an approved drug in the United States, where it is sold as a dietary supplement, usually in the form of tablets containing 40 mg of the extract. Recommended dosage for intermittent claudication is 120–160 mg daily in divided doses taken with meals. Daily dosage for cerebral circulatory disturbances, including early-stage dementia, is 240 mg in two or three divided doses. An initial six- to eight-week period is recommended to determine the efficacy of GBE. Seriously ill patients, such as those with stroke or advanced stages of dementia, should not replace physician care and evaluation with GBE self-medication.

Practically all of the scientific and clinical research on GBE has been carried out with extracts produced in Germany and designated either EGb 761 by the Dr. Willmar Schwabe GmbH & Co. or LI 1370 by Lichtwer Pharma GmbH. The bioequivalence of other GBE products has not been demonstrated.

Ginkgo leaves contain ginkgolic (anarcardic) acids (6-alkylsalicylic acids) that are allergenic and regarded as otherwise toxic but virtually eliminated in the process, which produces commercial proprietary extracts. The German Commission E limits the level of ginkgolic acids to a maximum of 5 ppm. Also present in dangerously high levels in unprocessed ginkgo seeds, ginkgotoxin (4'-O-methylpyridoxine), an antivitamin B_6 neurotoxin, is 99 percent inactivated by boiling; ginkgo leaves contain only minute quantities of the toxin.[51]

Other peripheral arterial circulatory disorders

The use of GBE for the treatment of intermittent claudication has already been discussed. This condition is a symptom of arteriosclerosis obliterans, which consists of segmented arteriosclerotic narrowing or obstruction of the lumen in the arteries supplying the limbs. It occurs primarily in men between the ages of fifty and seventy years. Risk factors include diabetes, hypertension, hypercholesterolemia, and smoking. Intermittent claudication denotes pain that develops usually in the legs during exercise and disappears when the patient rests. The pain is often described as a cramp or as severe fatigue of the exercising muscle. Because of the high frequency with which the femoral artery is involved, the most frequently affected muscles are those of the calf. Surgical treatment or balloon angioplasty is advisable when ischemia is present or if intermittent claudication seriously interferes with the patient's activities.[52]

Venous disorders

Varicose vein syndrome (chronic venous insufficiency)

Disorders of the veins most amenable to herbal therapy are those associated with varicosities. These are characterized by abnormally dilated, tortuous, readily visible, usually superficial veins, primarily of the extremities. The condition is often associated with aching discomfort or pain, depending on its severity and the presence of thrombus-associated inflammation, loss of endothelium, and edema. The superficial veins of the leg are most frequently affected because of the effects of gravity on venous pressure. Approximately 20 percent of adults develop varicose veins; the condition is more common in women, a reflection of the elevated venous pressure in the lower legs during pregnancy.[52]

It is now recognized that varicose vein syndrome is largely the result of the action of lysosomal enzymes that destroy the network of proteoglycans in the elastic tissue of the veins. This facilitates the passage of electrolytes, proteins, and water through the venous walls, thereby producing edema. The enzymes also act not only to reduce the strength of the vessel walls but also to cause them to dilate. The result is valvular incompetence, blood flow that tends to stagnate, and development of complications including thrombophlebitis and ulceration.[53]

Therapy for varicose vein syndrome has as its objectives:

- relief of existing edema;
- prevention of edema by reducing vessel permeability;
- reduction of blood stasis by increasing venous tonus;
- inhibition of the action of lysosomal enzymes; and
- neutralization of inflammatory and sclerotic processes.

In this connection, it must be noted that herbal treatment of varicose veins should not be expected to reverse changes in organic structures that have resulted from years of chronic varicosity. It may, however, provide some relief from the unpleasant symptoms by increasing capillary resistance, improving venous tonus, and even inhibiting the action of lysosomal enzymes.

Most of the phytomedicinals used to treat varicose vein syndrome are isolated chemicals or mixtures thereof, such as rutin, hesperidin, diosmin, and coumarin. These will not be discussed here.

Horse chestnut seed

By far the most effective plant drug employed for varicose vein syndrome is horse chestnut seed. The large, nearly globular, brown seeds of *Aesculus hippocastanum* L. (family Hippocastanaceae) or of closely related species such as the American horse chestnut or Ohio buckeye, *A. glabra* Willd., are probably known to every schoolchild because they have been widely used in children's games. Superstitious adults in many countries carry them in their pockets to prevent or to cure arthritis and rheumatism. Because of their attractive red, yellow, or white flower clusters, the relatively large trees that bear them are widely cultivated.

Horse chestnut seeds contain a complex mixture of triterpenoid saponin glycosides designated aescin. This may be fractionated into an easily crystallizable mixture known as β-aescin and water-soluble components referred to as α-aescin. Flavonoids, including quercetin and kaempferol, and the coumarin aesculetin are also present in the seed.[54]

Aescin has the ability to reduce lysosomal enzyme activity as much as 30 percent, apparently by stabilizing the cholesterol-containing membranes of the lysosomes and limiting release of the enzymes. The compound also restricts edema by reducing the transcapillary filtration of water and protein and has some beneficial diuretic effect. In addition, it has been shown to increase the tonus of the veins, thus improving return blood flow to the heart.[53] As can be seen from the earlier discussion, all of these actions would prove beneficial for the treatment of varicose veins and venous insufficiency.

Horse chestnut seeds are normally utilized in the form of an aqueous-alcoholic extract that is dried and adjusted to a uniform concentration of 16–21 percent triterpene glycosides, calculated as aescin. The efficacy of horse chestnut seed extract (HCSE) in reducing edema in patients with chronic venous insufficiency has been reported in several clinical trials.[54] A recent study compared the two modalities now commonly used to reduce edema: (1) standard mechanical compression with bandages and stockings is inconvenient, uncomfortable, and subject to poor patient compliance, and (2) the medicinal agent approach uses a venoactive substance, in this case HCSE, which exerts an inhibitory action on capillary protein permeability.

This randomized, partially blind, placebo-controlled clinical study involved 240 patients with a duration of twelve weeks. The medicinal agent group received 50 mg aescin twice daily, and the results were determined by measuring lower leg volume. Both those receiving compression therapy and HCSE therapy experienced a 25 percent reduction in mean edema volume, indicating that the two different modalities were equally effective.[55]

In the treatment of varicose veins, the initial oral dosage of HCSE is equivalent to 90–150 mg of aescin; following improvement, this may be reduced to 35–70 mg aescin equivalent daily. As of this writing, standardized horse chestnut extracts or dosage forms prepared from them are not commercially available in the United States. Extemporaneous preparation of a hydroalcoholic extract is possible, but standardization is difficult.

A number of ointments and liniments containing horse chestnut extract are available in Europe. Some of these are endorsed by the manufacturers not only for local application to superficial varicose veins but also for treatment of hemorrhoids. Because evidence regarding the transdermal absorption of aescin is lacking and hemorrhoids are related to both arterial and venous circulation, the efficacy of such preparations, particularly in the treatment of hemorrhoids, is doubtful.[56] Neither horse chestnut extract nor the aescin contained in it is recognized by the FDA as an effective ingredient in hemorrhoid preparations.

German Commission E has approved the use of horse chestnut seeds for the treatment of chronic venous insufficiency of various origins as well as pain and a feeling of heaviness in the legs. It is also recommended for varicose veins and postthrombotic syndrome. The herb is antiexudative and acts to increase tonus of the veins. Side effects are uncommon, but gastrointestinal irritation may occur.[57] Isolated cases of renal and hepatic toxicity as well as anaphylactic reactions have been reported following intravenous administration, but these appear to be exceptional.[58]

Butcher's broom

The rhizome and roots of *Ruscus aculeatus* L., a fairly common, short evergreen shrub of the family Liliaceae, have acquired some reputation recently as a useful treatment of varicose vein syndrome. Much less studied chemically and clinically than horse chestnut, the herb contains a mixture of steroidal saponins, particularly the modified cholesterol derivatives ruscogenin and neoruscogenin. Animal studies indicate that the activity of these saponins is in part linked to their stimulation of the postjunctional α_1- and α_2-adrenergic receptors of the smooth muscle cells of the vascular wall, which results in an increased tonus of the veins.[59]

In Europe, the herb is available in capsules or tablets containing about 300 mg each of a dried extract. Ointments and suppositories for the treatment of hemorrhoids are also available.[60] Butcher's broom capsules are available in the United States. Although toxicity has not been reported, much additional work must be carried out before the efficacy of butcher's broom can be established with certainty.

References

1. Johnson, J. A., and R. L. Lalonde. 1997. Congestive heart failure. In *Pharmacotherapy: A pathophysiologic approach,* ed. J. T. DiPiro, R. L. Talbert, G. C. Yee, G. R. Matake, B. G. Wells, and L. M. Posey, 219–256. Stamford, CT: Appleton & Lange.
2. Robbers, J. E., M. K. Speedie, and V. E. Tyler. 1996. *Pharmacognosy and pharmacobiotechnology,* 117–120. Baltimore, MD: Williams & Wilkins.
3. Bisset, N. G., ed. 1994. *Herbal drugs and phytopharmaceuticals,* English ed. (*Teedrogen,* M. Wichtl, ed.), 161–166. Boca Raton, FL: CRC Press.
4. Pöpping, S., H. Rose, I. Ionescu, Y. Fischer, and H. Kammermeier. 1995. *Arzneimittel-Forschung* 45 (II): 1157–1161.
5. Schüssler, M., J. Hölzl, A. F. E. Rump, and U. Fricke. 1995. *General Pharmacology* 26:1565–1570.
6. Hänsel, R. 1991. *Phytopharmaka,* 2nd ed., 25–40. Berlin: Springer–Verlag.
7. Schmidt, U., U. Kuhn, M. Ploch, and W.-D. Hübner. 1994. *Phytomedicine* 1:17–24.
8. Reuter, H. D. 1994. *Zeitschrift für Phytotherapie* 15:73–81.
9. Djumlija, L. C. 1994. *Australian Journal of Medicinal Herbalism* 6:37–42.
10. Schlegelmilch, R., and R. Heywood. 1994. *Journal of the American College of Toxicology* 13:103–111.
11. *Bundesanzeiger* (Cologne, Germany): January 3, 1984; May 5, 1988.
12. Schoen, F. J. 1994. Blood vessels. In *Robbins' pathologic basis of disease,* 5th ed., ed. R. S. Cotran, V. Kumar, and S. L. Robbins, 467–516. Philadelphia, PA: W. B. Saunders Co.

13. Agarwal, K. C. 1996. *Medicinal Research Reviews* 16:111–124.
14. Lawson, L. D. 1996. The composition and chemistry of garlic cloves and processed garlic. In *Garlic: The science and therapeutic application of* Allium sativum *L. and related species,* 2nd ed., ed. H. P. Koch and L. D. Lawson, 37–107. Baltimore, MD: Williams & Wilkins.
15. Kleijnen, J., P. Knipschild, and G. Ter Riet. 1989. *British Journal of Clinical Pharmacology* 28:535–544.
16. Lawson, L. D. 1993. Bioactive organosulfur compounds of garlic and garlic products. In *Human medicinal agents from plants,* ed. A. D. Kinghorn and M. F. Balandrin, 306–330. Washington, D.C.: American Chemical Society.
17. Pentz, R., and C.-P. Siegers. 1996. Garlic preparations: Methods of qualitative and quantitative assessment of their ingredients. In *Garlic: The science and therapeutic application of* Allium sativum *L. and related species,* 2nd ed., ed. H. P. Koch and L. D. Lawson, 109–134. Baltimore, MD: Williams & Wilkins.
18. Friedl, C. 1990. *Zeitschrift für Phytotherapie* 11:203.
19. Lawson, L. D., and B. G. Hughes. 1992. *Planta Medica* 58:345–350.
20. Lawson, L. D., Z.-Y. J. Wang, and B. G. Hughes. 1992. *Planta Medica* 57:363–370.
21. *Deutsche Apotheker Zeitung* 132:643–644 (1992).
22. Lawson, L. D., and Z. J. Wang. 2001. *Journal of Agriculture and Food Chemistry* 49:2592–2599.
23. Rahman, K., and G. M. Lowe. 2006. *Journal of Nutrition* 136:736S–740S.
24. Reuter, H. D. 1998. *Phytomedicine* 2:73–91.
25. Reuter, H. D., H. P. Koch, and L. D. Lawson. 1996. Therapeutic effects and applications of garlic and its preparations. In *Garlic: The science and therapeutic application of* Allium sativum *L. and related species,* 2nd ed., ed. H. P. Koch and L. D. Lawson, 135–212. Baltimore, MD: Williams & Wilkins.
26. *Bundesanzeiger* (Cologne, Germany): July 6, 1988.
27. Imai, K., and K. Nakachi. 1995. *British Medical Journal* 310:693–696.
28. Luc, M., K. Kannar, M. L. Wahlqvist, and R. C. O'Brien. 1997. *Lancet* 349:360–361.
29. Fink-Gemmels, J., J. Dresel, and L. Leistner. 1991. *Fleischwirtschaft* 71:329–331.
30. Walker, L. A. 1997. Drug topics. Natural Products Update Supplement, June: 8–10.
31. Robbers, J. E., M. K. Speedie, and V. E. Tyler. 1996. *Pharmacognosy and pharmacobiotechnology,* 110–112. Baltimore, MD: Williams & Wilkins.
32. McEvoy, G. K., ed. 1997. *The American hospital formulary service: Drug information,* 1349–1357. Bethesda, MD: American Society of Health-Systems Pharmacists.
33. Heber, D., I. Yip, J. Ashley, D. A. Elashoff, R. M. Elashoff, and V. L. Go. 1999. *American Journal of Clinical Nutrition* 69:231–236.
34. Pharmanex, http://www.pharmanex.com (September 2001).
35. Dumoff, A. 2001. *Alternative and Complementary Therapies* October: 310–314.
36. Talbert, R. L. 1997. Peripheral vascular disease. In *Pharmacotherapy: A pathophysiologic approach,* ed. J. T. DiPiro, R. L. Talbert, G. C. Yee, G. R. Matake, B. G. Wells, and L. M. Posey, 491–508. Stamford, CT: Appleton & Lange.
37. Almada, A. http://www.ffnmag.com/ASP/articleDisplay.asp?strArticleId= 403&strSite=FFNSite (October 2006).

38. Gouni-Berthold, I., and H. K. Berthold. 2002. *American Heart Journal* 143:356–365.
39. Lin, Y. 2004. *Metabolism* 53:1309–1314.
40. Reiner, Z. 2005. *Clinical Drug Investigation* 25:701–707.
41. Greyling, A., C. De Witt, and W. Oosthuizen. 2006. *British Journal of Nutrition* 95:968–975.
42. Berthold, H. K. 2006. *Journal of the American Medical Association* 295:2262–2269.
43. Bradberry, J. C. 1997. Stroke. In *Pharmacotherapy: A pathophysiologic approach,* ed. J. T. DiPiro, R. L. Talbert, G. C. Yee, G. R. Matake, B. G. Wells, and L. M. Posey, 435–458. Stamford, CT: Appleton & Lange.
44. Kumar, V., R. S. Cotran, and S. L. Robbins. 1997. *Basic pathology,* 6th ed., 3–24. Philadelphia, PA: W. B. Saunders Co.
45. Blumenthal, M. 1996. *HerbalGram* 38:58–59.
46. Del Tredici, P. 1991. *Arnoldia* 51:2–15.
47. Hänsel, R. *Phytopharmaka*, 2nd ed., 59–72. Berlin: Springer–Verlag.
48. Campbell, W. B., and P. V. Halushka. 1996. Lipid-derived autacoids. In *Goodman and Gilman's the pharmacological basis of therapeutics,* 9th ed., ed. J. G. Hardman, L. E. Limbird, P. B. Molinoff, and R. W. Puddon, 601–616. New York: McGraw–Hill.
49. Meisel, C., A. Johne, and I. Roots. 2003. *Atherosclerosis* 165:367 (letter).
50. Koch, E. 2005. *Phytomedicine* 12:10–16.
51. Bent, S., H. Goldberg, A. Padula, and A. L. Avins. 2005. *Journal of General Internal Medicine* 20 (7): 657–661.
52. Kanowski, S., W. M. Herrmann, K. Stephan, W. Wierich, and R. Hörr. 1996. *Pharmacopsychiatry* 29:47–56.
53. LeBars, P. L., M. M. Katz, N. Berman, T. M. Itil, A. M. Freedman, and A. F. Schatzberg. 1997. *Journal of the American Medical Association* 278:1327–1332.
54. Schneider, B. 1992. *Arzneimittel Forschung* 42:428–436.
55. Grews, W. D. Jr., D. W. Harrison, M. L. Griffin, K. D. Falwell, T. Crist, L. Longest, L. Hehemann, and S. T. Rey. 2005. *HerbalGram* 67:43–62.
56. Elsabagh, S., D. E. Hartley, O. Ali, E. M. Williamson, and S. E. File. 2005. *Psychopharmacology* 179:437–446.
57. Roisenblatt, M., and J. Mindel. 1997. *New England Journal of Medicine* 336:1108.
58. Huh, H., and E. J. Staba. 1992. *Journal of Herbs, Spices & Medicinal Plants* 1:92–124.
59. Schoen, F. J. 1994. Blood vessels. In *Robbins' pathologic basis of disease,* 5th ed., ed. R. S. Cotran, V. Kumar, and S. L. Robbins, 467–516. Philadelphia, PA: W. B. Saunders Co.
60. Haas, H. 1991. *Arzneipflanzenkunde,* 31–40. Mannheim, Germany: B. I Wissenschaftsverlag.
61. Newall, C. A., L. A. Anderson, and J. D. Phillipson. 1996. *Herbal medicines: A guide for health-care professionals,* 166–167. London: The Pharmaceutical Press.
62. Diehm, C., H. J. Trampisch, S. Lang, and C. Schmidt. 1996. *Lancet* 347:292–294.
63. Weiss, R. F. 1988. *Herbal medicine,* 187. Gothenburg, Sweden: AB Arcanum.
64. *Bundesanzeiger* (Cologne, Germany): December 5, 1984.

65. Vogel, G. 1989. *Zeitschrift für Phytotherapie* 10:102–106.
66. Bouskela, E., F. Z. G. A. Cyrino, and G. Marcelon. 1994. *Journal of Cardiovascular Pharmacology* 24:165–170.
67. Braun, H., and D. Frohne, D. 1987. *Heilpflanzen-Lexicon für Ärtze and Apotheker,* 212–213. Stuttgart: Gustav Fischer Verlag.

chapter seven

Nervous system disorders

Anxiety and sleep disorders

Anxiety is apprehension, tension, or uneasiness that stems from external stress that may be associated with the trials and tribulations of everyday living or, particularly, as a result of a traumatic life event (e.g., bereavement, financial losses, or divorce), or it may be devoid of any apparent cause. Anxiety is usually a normal response experienced by all people, and a certain amount of anxiety in one's life is a necessary corollary of need-fulfilling activity, coping, and personal growth.

On the other hand, when anxiety becomes disproportionate to the causal stimulus or when no stimulus can be identified, it becomes disruptive, and the symptoms can interfere with a person's ability to function effectively. At this point, the anxiety is considered pathologic and may be classified as an anxiety disorder. Approximately 7 percent of adults in the United States are affected by one of these disorders as classified by the American Psychiatric Association. They include, among others, panic disorder with and without agoraphobia, generalized anxiety disorder, obsessive–compulsive disorder, and post-traumatic stress disorder.[1,2] Therapy is based principally on the type and degree of anxiety and should be conducted under the care of a physician.

Insomnia, the inability to attain restful sleep in adequate amounts, is often a transient response to the anxiety produced by stressful situations. About 30 percent of the American population experiences insomnia in some form over the course of a year, and it is a common complaint among the elderly. Insomnia is an inclusive term that can include problems in falling asleep (longer than thirty minutes), frequent awakening through the night with difficulty in immediately returning to sleep, and early morning final awakening resulting in a total sleep time of fewer than six hours. It can be transient, lasting for less than one week, or chronic if it lasts longer. The latter category may also be symptomatic of more serious physiologic or psychological conditions or the use of various drugs, including alcohol.[3]

Anxiety and insomnia are amenable to treatment with drugs that exert a depressant effect on the central nervous system. In many cases, the same central nervous system (CNS) depressants are used to treat both

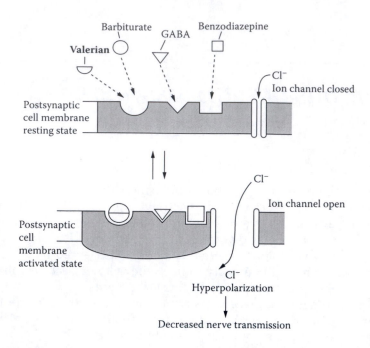

Figure 7.1 A generalized model for the mechanism of action of some anxiolytic and sleep aid drugs at the $GABA_A$ receptor macromolecular complex of the postsynaptic neuronal membrane. Drugs do not bind at the GABA recognition protein but, rather, at separate recognition proteins that enhance the effect of GABA.

conditions; however, a larger dose is customarily employed to induce sleep. The agents used are referred to by a number of names, including sleep aids, sedatives, hypnotics, soporifics, antianxiety agents, anxiolytics, calmatives, and minor tranquilizers. All of these terms are more or less synonymous, although, as noted, the degree of response obtained is dose dependent.

The mechanism of action for many of these drugs involves gamma-aminobutyric acid (GABA), which is the major inhibitory neurotransmitter in the CNS. For example, benzodiazepines, an important chemical group of drugs for treating anxiety and insomnia, potentiate GABAergic neurotransmission at all levels of the CNS. Their action, illustrated in Figure 7.1, involves an interaction with a $GABA_A$ receptor–chloride ion channel macromolecular complex, causing an increase in the frequency of chloride ion channel opening in the postsynaptic membrane of the neuron. This, in turn, leads to membrane hyperpolarization and a decrease in the firing rate of critical neurons in many regions of the brain. Barbiturates also facilitate the actions of GABA; however, in contrast to benzodiazepines, they increase the duration of the GABA-gated channel openings.[4] Most of the prescription drugs in these categories involve some risk of overdose, tolerance, habituation, and addiction.

The herbs commonly used for their sedative effects do not suffer these drawbacks, but neither do they possess the degree of activity shown by the prescription drugs. In fact, the action of many traditional plant hypnotics is so slight as to remain uncertain.

Herbal remedies for anxiety and sleep disorders

Valerian

One of the most effective herbal remedies of the entire group is valerian. The dried rhizome and roots of a tall perennial herb, *Valeriana officinalis* L. (family Valerianaceae), have enjoyed a considerable reputation as an anxiolytic agent and sleep aid for more than a thousand years. The related species *V. wallichii* DC syn. *V. jatamansi* Jones (Indian valerian), *V. edulis* Nutt. syn. *V. mexicana* DC (Mexican valerian), and *V. fauriei* Briquet (Japanese valerian) are also used medicinally.[5] The roots of *V. officinalis* (common or "officinal" valerian) contain from 0.3 to 0.7 percent of an unpleasant smelling volatile oil containing monoterpenes (mainly bornyl acetate and bornyl isovalerate) and the sesquiterpene derivatives valerenal, valeranone, valerenic acid, and acetoxyvalerenic acid.

Valerenic acids occur only in the valerian species *officinalis*; hydroxyvalerenic acid, the product of hydrolysis of acetoxyvalerenic acid, may be found at very low levels. Also present is 0.5–2 percent of a mixture of lipophilic iridoid principles known as valepotriates (valerian epoxy triesters).[6] These bicyclic monoterpenes are quite unstable and occur only in the fresh plant material or in that dried at temperatures under 40°C. In addition, various sugars, amino acids, free fatty acids, and aromatic acids have been isolated from the drug.[7] A monograph on valerian was produced by the American Herbal Pharmacopoeia in 1999.[8]

Identity of the active principles of valerian has been a subject of controversy for many years. Initially, the calmative effect was attributed to the volatile oil; indeed, this kind of activity was long associated with most herbs containing oils with disagreeable odors. Then, beginning in 1966 with the isolation of the valepotriates, the property was attributed to them for a twenty-year period. This was done in spite of the fact that they were highly unstable and were contained in most valerian preparations only in small amounts. Finally, in 1988, Krieglstein and Grusla showed that although valerian did produce CNS depression, the tested valepotriates, the sesquiterpenes valerenic acid or valeranone, and the volatile oil itself displayed no such effects in rats.[9]

Although the active principles of valerian remain unidentified, it appears that a combination of volatile oil, valepotriates, and possibly certain water-soluble constituents is likely involved. Its aqueous extract has been regarded as mostly responsible for its sleep-inducing effect, while its volatile oil (mainly valerenic acid) and the valepotriates (through their degradation products, such as valtroxal and baldrinals) are likely major contributors to its sedative effects. Current data also indicate the valepotriates to be responsible for the antianxiety or tranquilizing effect of valerian. No serious toxic side effects associated with clinical use of valerian have been reported.

As with other anxiolytic and sedative-hypnotic drugs, laboratory studies support a mechanism of action in the CNS (see Figure 7.1). Neuropharmacological studies in mice using an ethanol extract of valerian demonstrated that it prolonged the effects of a barbiturate (thiopental) and was effective as an anticonvulsant against picrotoxin, which is a $GABA_A$ receptor antagonist. The site for valerian action may be the same as or near the barbiturate recognition site because barbiturates potentiate GABA by an interaction at the picrotoxin site of the $GABA_A$ receptor complex.[10]

A number of clinical investigations have demonstrated the sedative effect of valerian and its effectiveness as an anxiolytic and mild hypnotic in the treatment of insomnia. It has a beneficial effect on several sleep-disorder parameters such as sleep latency, final wake time after sleep, frequency of waking, nighttime motor activity, inner restlessness, and quality of sleep.[11] The results of clinical study mainly addressing sleep disorders are summarized in considerable detail in the second edition of *Botanical Medicines*.[12] Based on these generally favorable results, the German Commission E has approved valerian as a calmative and sleep-promoting agent useful in treating states of unrest and anxiety-produced sleep disturbances.[13]

The herb may be administered several times daily as a tea prepared from 2–3 g (1 teaspoonful) of the dried rhizome and roots. Equivalent amounts (0.5–1 teaspoonful) of a tincture or extract may also be employed. Valerian is also used externally in the form of a calmative bath (100 g per tub of water), but evidence supporting its effectiveness by this route is much less substantial. In the preparation of any of these dosage forms, the use of fresh or recently and carefully dried (temperatures under 40°C) herb is most likely to yield satisfactory results. Side effects and contraindications to the use of valerian have not been reported; however, in vitro cytotoxicity and mutagenicity have been reported for the valepotriates. The clinical significance is probably of no concern because valepotriates decompose rapidly in the stored drug and also are not readily absorbed.

Recently, an attempted suicide was reported when 20 g (ten times the recommended dosage of 2 g) of powdered valerian root was ingested. After thirty minutes, the patient complained of fatigue, abdominal cramping, chest tightness, lightheadedness, and feet and hand tremors. All other tests, including liver function, were normal and all symptoms were resolved within twenty-four hours.[14]

Kava

Kava is the dried roots of *Piper methysticum* Forst., family Piperaceae. Known also as kava-kava, the plant is a large shrub widely cultivated in Oceania. Its underground parts have been extensively used by the indigenous people of these Pacific islands in the preparation of an intoxicating beverage. Kava beverage drinking predates recorded history and has acquired important significance from a social and ceremonial standpoint among the various island cultures. An overview of kava preparation and ceremonial use as well as kava chemistry and pharmacology has been presented in detail by Singh and Blumenthal.[15] Kava was ritually prepared by young men or women who chewed cut pieces of the scraped root, spitting the pieces into a bowl to which coconut milk was added to make an infusion. However, for sanitary reasons, pressures by colonizing governments and missionary influence have led to the abandonment of the chewing technique in favor of pounding or grating the root.

Kava bars offer the beverage on various Pacific islands, where the muddy-looking liquid is rapidly drunk (not sipped) from coconut shells. It first causes a numbing and astringent effect in the mouth, followed by a relaxed sociable state in which fatigue and anxiety are lessened. Eventually, a deep restful sleep ensues from which the user awakes refreshed and without hangover. Excessive consumption can lead to dizziness and stupefaction,[16] and use over a prolonged period (a few months) can lead to kava dermopathy, which is characterized by a reversible ichthyosiform scaly skin.[17]

The CNS activity of kava is due to a group of resinous compounds called kavalactones, kavapyrones, or styrylpyrones. Roots of good quality contain between 5.5 and 8.3 percent of kavalactones, including the major components kawain, dihydrokawain, and methysticin, as well as the minor compounds yangonin, desmethoxyyangonin, and dihydromethysticin.[18]

Animal studies have demonstrated the ability of kava extracts or purified kavalactones to induce sleep and produce muscle relaxation and analgesia as well as anticonvulsive protection against strychnine and electroshocks.[19] At low doses, kavalactones resemble the pharmacologic action of benzodiazepines, although they apparently do not bind at either GABA or benzodiazepine binding sites in rat or mouse brain membranes.[20] Because of the diverse range of pharmacological activities shown by kava, kavalactones may act nonselectively and remain in the lipid membranes to cause modifications to a variety of receptor domains rather than interacting with any specific receptor binding site. There is evidence that the spasmolytic, analgesic, and anticonvulsant activity of kavalactones might be explained by inhibition of voltage-dependent Na^+ channels in the brain[21]; however, the question remains whether this inhibition is also responsible for the anxiolytic and hypnotic action.

Several relatively short-term clinical studies have provided favorable evidence that kava is effective in treating anxiety and insomnia[22]; however, the first long-term clinical trial investigating safety and efficacy of kava in anxiety patients was only recently published. In this study, 101 outpatients with anxiety of nonpsychotic origin, determined on the basis of the American Psychiatric Association anxiety disorder classification, were included in a twenty-five-week multicenter, randomized, placebo-controlled, double-blind trial using a lipophilic kava extract standardized to contain 70 percent kavalactones. Patients were administered either 100 mg of extract (70 mg kavalactones) or placebo three times daily, and the main outcome criterion utilized the Hamilton Anxiety Scale.

Adverse effects were rare and distributed evenly in both groups, and there were no problems with tolerance, dependence, or withdrawal symptoms—adverse effects often associated with long-term benzodiazepine use. The long-term efficacy (after the eighth week of treatment) of the kava extract was superior to that of placebo. In addition, kava extract was not associated with depressed cognitive function, drowsiness, or impairment in mental reaction time, problems found in the side-effect profile of benzodiazepines.[23]

The German Commission E has approved kava for conditions of nervous anxiety, stress, and restlessness; the recommended dosage is 70 mg of kavalactones, usually in the form of a standardized extract two to three times daily.[24] As a sleep aid, 180–210 mg of kavalactones can be taken one hour before bedtime. It is important to note that ethanol and other CNS depressants can potentiate the effects of kava; therefore, they should not be taken concomitantly. In this regard, a drug interaction between kava and the benzodiazapine alprazolam that caused a semicomatose state in a patient has been reported.[25]

In June 2002, the German Federal Institute of Drugs and Medical Devices decided to withdraw all drug registrations for all products containing kava, on account of reports of liver toxicity associated with the herb. However, in May 2005, the German government repealed the ban on kava products, pending evaluation of new data. An article by Anke and Ramzan in 2004 reviewed the kava hapatotoxicity controversy and summarized the major theories advanced to explain the case reports of liver failure but reached no satisfactory conclusion.[26] An intriguing hypothesis has been recently proposed implicating a kava metabolite with the potential for reacting with glutathione in the liver, supported by an adduct identified in urine.[27] Nonetheless, German herbal medicine experts, including former members of Commission E and officials of the Society of Medicinal Plant Research, regard kava extracts and kavalactones as "quite safe."[28]

Passion flower

A 1986 survey of herbal sedatives in Britain revealed that the most popular, based on its incorporation in the largest number of proprietary preparations, was passion flower.[29] Consisting of the dried flowering and fruiting top of a perennial climbing vine, *Passiflora incarnata* L. (family Passifloraceae), passion flower is, in spite of its popularity, a relatively unproven anxiolytic agent. Its principal constituents include up to 2.5 percent flavonoids, especially vitexin, as well as coumarin, umbelliferone, and 0.05 percent maltol. Harmala-type indole alkaloids, including harman, harmine, harmaline, and harmalol, have been reported in subtherapeutic amounts (up to 0.01 percent).[30]

Administered intraperitoneally to rats, passion flower extract significantly prolonged sleeping time and affected locomotor activity. These and related activities could not be attributed to the flavonoids or the alkaloids in the extract.[31] Clinical studies are required to verify such activity in human beings.

In 1978, the FDA prohibited the use of passion flower in OTC sedative preparations on the grounds that it had not been proven safe and effective.[32] The German Commission E, however, has authorized its use in the treatment of nervous unrest, citing its ability to bring about a reduction in mobility in animal experiments.[33] The usual daily dose is 4–8 g (3–6 teaspoonfuls) taken as a tea in divided doses. Side effects and contraindications have not been reported.

Hops

Another popular herbal sleep aid is hops, the dried strobile with its glandular trichomes of *Humulus lupulus* L., family Cannabinaceae. The herb contains about 15–30 percent of a resin that accounts for its bitter taste and for its well-established use as a preservative in the brewing of beer. On the other hand, the sedative properties of hops are not well established, and much misinformation exists in the literature.

Hänsel and Wagener found that hop resin had no CNS depressant effect when taken orally.[34] However, when the herb is stored, bitter principles such as humulone, lupulone, and colupulone undergo auto-oxidation to produce a C_5-alcohol designated 2-methyl-3-buten-2-ol or, more simply, methylbutenol. A hops sample stored for two years contains about 0.04 percent methylbutenol. This volatile compound does produce CNS-depressant effects when inhaled, and this has been postulated to account for the soporific effects of hop-filled pillows. However, a single effective dose of methylbutenol would require all of that compound contained in 150 g of hops.[35]

On the other hand, it has been demonstrated that colupulone is a potent inducer of mouse hepatic cytochrome P450, and it has been suggested that 2-methyl-3-buten-2-ol is formed in vivo through oxidation catalyzed by the P450 enzyme.[36] However, it must be concluded that both the CNS-depressant activity and the identity of any active sedative principles in hop extracts or preparations are questionable at this time.

Bacopa

The whole plant of *Bacopa monnieri* (L.) Penn; synonym *B. monniera* (L.) Wettst. (family Scrophulariaceae) is the source of the Ayurvedic drug Brahmi, usually dried and powdered. Modern scientific investigations have generally employed ethanolic extracts of macerated plant material, filtered and evaporated to dryness under vacuum.

Bacopa saponins (dammarane-type tetracyclic triterpenoids) are widely regarded as the plant's active principles responsible for its scientifically supported therapeutic properties, Bacoside A, the putative main bioactive component, has been determined to be a mixture of saponins, with bacoside A3, bacopaside II, jujubogenin isomer of bacopasaponin C, and bacopasaponin C the major constituents. Bacoside B is also a mixture of saponins.[37]

A syrup made from Brahmi was administered in two divided doses totaling 30 mL (equivalent to a daily dose of 12 g of dried plant) for four weeks to thirty-five adult patients suffering from anxiety neurosis. The *Bacopa* preparation significantly reduced anxiety as well as improved mental performance, concentration and immediate memory; also noted was a reduction in mental fatigue, a general feeling of well-being, improved sleep quality and appetite, and increased body weight. Side effects were minimal and clinically insignificant.[38] An experimental study, using a rat model of clinical anxiety, was conducted in 1998. A *Bacopa* extract, standardized to 25 percent bacoside A, exerted anxiolytic activity comparable to Lorazepam, a common benzodiazepine anxiolytic drug, but without the amnesic side effect associated with the pharmaceutical; rather, it had a memory-enhancing influence.[39] Also, the potential of *Bacopa* as an anti-stress agent was assessed in rats by judging the effect of a standardized *B. monnieri* extract on both acute (AS) and chronic (CS) stress, using *Panax quinquefolius* root powder as a standard. Pretreatment with a low dose of *B. monnieri* extract (40 mg/kg) significantly reversed AS-induced changes in ulcer index and plasma aspartate aminotransferase (AST). Pretreatment with a higher dose significantly reversed CS-induced changes in ulcer index, adrenal gland weight, creatine kinase, and AST.[40]

Evidence for *Bacopa*'s beneficial effect on cognitive function; juvenile IQ/intellectual function, including attention deficit hyperactivity disorder (ADHD); epilepsy; irritable bowel syndrome; and other conditions was reviewed in a published monograph.[41] The usual dose for cognitive effect is 2 × 150 mg/d of an ethanolic extract standardized to at least 50 percent combined bacosides for three months[42]; acute administration of *B. monnieri* extract had no effect on cognitive function.[43] *Bacopa* preparations have been used safely in Ayurvedic medicine for several hundred years.[41] Concentrated bacosides administered to healthy males in single doses (20–300 mg) or in multiple doses (100–200 mg) daily for four weeks were well tolerated and devoid of adverse effects.[44]

Gotu kola

Gotu kola (meaning "cup-shaped leaf" in Sinhalese, the official language of Sri Lanka) is an aromatic creeping perennial herb indigenous to sub-tropical and tropical climates that thrives in the high humidity of water banks, swampy areas, and moist, sunny, rocky places. The plant, *Centella asiatica* (L.) Urban, synonym *Hydrocotyle asiatica* L. (family Apiaceae), is termed *Brahmi* in India, like *Bacopa monnieri*, and also *Manduka parni*. Traditionally, extracts of the plant have been used topically and internally for wound healing and treating leprosy. Since the nineteenth century, the plant and its extracts have been recommended in India for skin condi-tions associated with eczema, lupus, and psoriasis and for varicose ulcers. Various other regional traditional applications have been recorded.[45] *C. asiatica* was also regarded as particularly beneficial in female conditions such as amenorrhea and disease of the genital and urinary systems.[46]

The medicinal parts of the plant are the dried aerial parts and the fresh and dried leaves and stem. The active principles are viewed as triter-penoids, mainly asiatic acid, madecassic acid, asiaticoside, and medacas-soside.[47] Unrelated to cola or kola nut (*Cola nitida, C. acuminata*), gotu kola contains no caffeine.

Extracts of fresh and dried leaves of *C. asiatica* have been shown to possess mild tranquilizing, antistress, and antianxiety ability, appar-ently by enhancing cholinergic mechanisms.[48] An evaluation of the influ-ence on the CNS of an aqueous extract of *C. asiatica* was conducted in mice. Administered intraperitoneally, the extract decreased spontaneous motor activity and delayed pentylenetetrazole (PTZ)-induced convul-sions to a degree comparable to the effect of diazepam; it also potenti-ated pentabarbitone-induced sleep but did not affect immobility time in swimming tests.[49] An ethanolic extract of *C. asiatica* was found to exhibit significant antistress activity in a variety of stress tests on rats, also using diazepam as control. Antidepressant activity was also indicated and an analgesic effect demonstrated, along with a reduction in the adrenocorti-cal response.[50] Studies with a hydroalcoholic extract of *C. asiatica* leaves have confirmed the plant's anxiolytic/sedative effect, potentiating in rats the hypnotic effect of phenobarbitone, as well as anticonvulsant activity against PTZ, as previously observed for the aqueous extract in mice.[49]

A double-blind, placebo-controlled study conducted at the Royal Ottawa Hospital and the Department of Psychiatry, University of Ottawa, Ontario, Canada, examined the effects of gotu kola on acoustic startle response (ASR), reported to be attenuated by gotu kola in rats. Forty healthy subjects (twenty-one men and nineteen women), aged eighteen to forty-five years, were administered orally a single 12-g dose of pow-dered dried leaf mixed in 300 mL of grape juice (*n* = 20) or 300 mL of plain grape juice (*n* = 20) matched for color, taste, and smell. ASR, heart rate, blood pressure, and mood data were recorded at baseline and at 30, 60, 90, and 120 minutes after ingestion. The startle response was statistically significantly lower thirty and sixty minutes after ingestion of gotu kola as compared to placebo, with an affect size of 0.48 and 0.77, respectively, supporting anxiolytic activity. There was no difference between placebo and treatment at baseline and at later times, as was the case with heart rate, blood pressure, and mood response, except for self-rated energy level.[51]

The effects of *C. asiatica* on learning and memory have also been explored. The effect of different extracts on cognition and markers of oxidative stress in rats has been studied. Only the aqueous extract of whole plant (200 mg/kg for fourteen days) showed an improvement in learning and memory in both shuttle box and step-through paradigms; the indication was that an antioxidant mechanism was involved. One interesting double-blinded clinical trial in India in 1977 observed a significant enhancement of mental abilities in thirty "developmentally disabled" children (seven girls and twenty-three boys, aged seven to eighteen years) treated with a gotu kola preparation. After six months, the children exhibited better overall adjustment, were more attentive, and were better able to concentrate on assigned tasks. Unfortunately, the treatment is described only as "one tablet (0.5 g) a day for six months."[53] The potential of gotu kola in this area ought to be further explored.

l-Tryptophan

l-Tryptophan is not an herb but, rather, a natural amino acid that occurs in concentrations of 1–2 percent in many plant and animal proteins. It is produced in quantity by fermentation utilizing selected bacteria. *l*-Tryptophan is an essential amino acid that must be obtained from exogenous sources; the minimal daily requirement for an adult is 3 mg (30 percent) per kilogram of body weight. For a 180-lb person, this is equivalent to about 0.25 g.[54]

Studies have shown that 1-g doses of *l*-Tryptophan reduce sleep latency by increasing subjective "sleepiness" and also decrease waking time.[55] The amino acid is thought to function by bringing about an increase in serotonin in certain brain cells, thus inducing sleep but also assisting in the treatment of mental depression.[56] Although never approved as a drug, capsules and tablets of the amino acid in amounts ranging from 100 to 667 mg were widely sold as sleep aids in health food stores until 1989.

In that year, cases of a serious blood disorder, eosinophilia-myalgia syndrome (EMS), began to occur in otherwise healthy individuals who had consumed quantities of *l*-Tryptophan. The FDA ordered the recall of manufactured over-the-counter *l*-Tryptophan in any form at all dosage levels, but the product had already caused more than fifteen hundred cases of EMS and thirty-seven deaths.[57] Eventually, all of the suspect amino acid was traced to a single manufacturer, Showa Denko K. K. in Japan. That organization had used a new genetically engineered bacterium to produce the *l*-Tryptophan and had also modified the customary purification procedure. Suspicion regarding the cause of EMS then shifted away from the *l*-Tryptophan to contaminants produced, or at least not removed, during the fermentation process. One contaminant has now been identified as 1,1i-ethylidenebis(*l*-Tryptophan); another was found to be 3-(phenylamino)-l-alanine. The latter compound is similar to a contaminant in industrial rapeseed oil that was responsible for an outbreak of EMS in Spain in 1981 when the impure oil was used for culinary purposes.[40] In total, seventeen contaminants have been determined in an EMS-implicated batch of *l*-Tryptophan, and one or more of these chemicals may trigger EMS by an undefined mechanism.[58]

Some uncertainties still remain about the ability of *l*-Tryptophan manufacturers to produce a pure amino acid free of potentially toxic contaminants. The precise contaminant causing the toxicity is still unknown, and this has been complicated by the fact that up to 5 percent of EMS cases have not been linked to contaminated *l*-Tryptophan, suggesting that large doses of noncontaminated *l*-Tryptophan may also cause EMS due to abnormalities in metabolism.[59] Consequently, the FDA has not yet authorized the sale of *l*-Tryptophan as a dietary supplement. Certainly, additional research into the safety and efficacy of pure *l*-Tryptophan as a potentially valuable therapeutic agent is needed. Until the time when safety of the marketed product can be ensured, the consumption of manufactured *l*-Tryptophan in any form must be avoided.

Melatonin

Melatonin (*N*-acetyl-5-methoxytryptamine) is a hormone produced in the pineal gland of all mammalian species and for commercial purposes by chemical synthesis. In humans, melatonin synthesis and secretion are increased during the dark period of the day and maintained at low levels during the daylight hours. This remarkable diurnal variation in synthesis is controlled by the eyes' perception of light and darkness, and the mechanism is illustrated in Figure 7.2. The retina of the eye is connected to the pineal gland via a multineuronal pathway that involves the suprachiasmatic nucleus of the hypothalamus and then by preganglionic neurons in the upper thoracic spinal cord that innervate the superior cervical ganglia—the site of origin of the postganglionic sympathetic neurons that secrete norepinephrine. Norepinephrine acts via β-adrenergic receptors in the pineal gland to increase intracellular cyclic AMP, and the cyclic AMP in turn causes a marked increase in *N*-acetyltransferase activity, resulting in the synthesis of melatonin.[60]

In humans, melatonin has both a sleep-promoting effect and a role in synchronizing the sleep/wake cycle with the circadian dark/light rhythm.

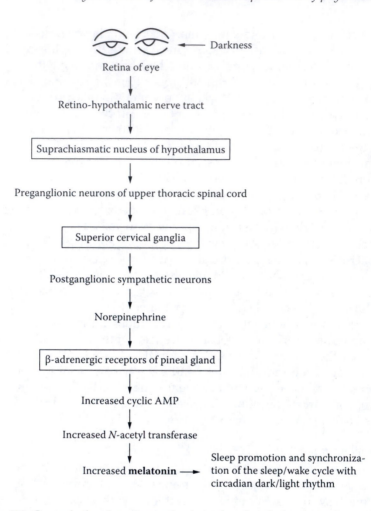

Figure 7.2 Control of melatonin synthesis in the pineal gland via the retina and hypothalamus.

Nocturnal melatonin levels are highest in early childhood, drop through adolescence, remain fairly constant until late adult life, and then show a decline. This impairment in melatonin production may contribute to the increased frequency of sleep disorders in the elderly, and melatonin replacement therapy may be effective in improving sleep quality in this population. A limited study investigating the effect of a 2-mg controlled-release formulation of melatonin on sleep quality over three weeks in twenty-three elderly subjects who complained of insomnia found that sleep efficiency was significantly greater after melatonin than after placebo and wake time after sleep onset was significantly shorter.[61] In a separate study, subjects received low doses (0.3 or 1 mg) of melatonin in the evening at 6, 8, or 9 p.m. Either dose given at any of the three time points decreased sleep onset latency and latency to stage 2 sleep. Melatonin did not suppress REM sleep or produce hangover effects on the morning after treatment.[62]

In patients with difficulty maintaining sleep, low doses of melatonin may not produce sufficient blood concentrations to maintain slumber. A 2-mg oral melatonin dose in young adults produced peak levels approximately ten times higher than nighttime physiological levels; however, it did not alter endogenous production, and the elevated level lasted for only three to four hours. To maintain effective serum concentrations of melatonin throughout the night, high doses, repeated low doses during the night, or sustained release formulations may be needed.[63]

Administration of melatonin during the day, when melatonin is not normally produced at detectable levels, has been reported to decrease body temperature and suggests that the hypnotic effects of melatonin are mediated by a decrease in body core temperature. A recent study shows that a β-adrenergic blocking drug, atenolol, can suppress endogenous nocturnal melatonin in young men to low levels typical of the elderly. When the quality of sleep in these subjects was studied, it was found they had increased nighttime wakefulness with difficulty in returning to sleep, decreased REM sleep and deep sleep, and an increase in mean nocturnal core temperature. Melatonin may, therefore, be useful as a treatment for sleep disorders related to elevated core temperature, such as those suffered by the elderly and shift workers.[64]

The ability of melatonin to modulate circadian rhythms has prompted several studies investigating its use in the prevention of jet lag. In a double-blind, placebo-controlled trial in international flight attendants, 5 mg of melatonin taken daily for five days, beginning on the day of departure, was shown to decrease jet lag and sleep disturbance.[65] Ideally, melatonin should be administered at bedtime or after boarding the airplane because subjects may experience drowsiness starting within thirty minutes after ingestion and lasting for about one hour.

Adverse effects from the use of melatonin appear to be minimal. Headache and transient depression have been reported, and because of possible next-day hangover effects, patients should be cautioned not to drive or operate machinery after taking the drug. Effects associated with long-term administration have not been studied. The FDA has not received any reports of adverse reactions to melatonin yet cautions that consumers who take melatonin are doing so without any assurance that it is safe or that it will have any beneficial effect. Until further studies have been completed, melatonin should be avoided by pregnant women, nursing mothers, and children (because they naturally produce high levels of melatonin) and psychiatric patients (because melatonin may aggravate depressive symptoms).[66]

The use of melatonin for sleep disorders has been obfuscated by a host of best-selling books and popular magazine articles touting it as an antiaging agent, oral contraceptive, cancer preventive, antioxidant, and immunostimulant, to name a few. Clinical trials in humans to support these claims are lacking, and large-scale double-blind studies are needed before melatonin can be recommended for any of these therapeutic indications.

Melatonin has been detected in feverfew [*Tanacetum parthenium* (L.) Schultz Bip.] at levels ranging from 1.37 to 2.45 µg/g of dried leaf, St. John's wort (*Hypericum perforatum* L.) at 4.39 µg/g of dried flowers, and Baikal or Chinese skullcap (*Scutellaria baicalensis* Georgi) at 7.11 µg/g of dried leaves.[67] However, it has not been established whether these levels of melatonin are of clinical significance in medication with commercial products derived from these plants.

Depression

Depression—the inability to experience pleasure or happiness, usually accompanied by feelings of helplessness and lack of self-worth—is one of the most common of the psychiatric disorders. It is a so-called affective disorder, meaning that it causes changes in emotions, feelings, or mental state. It is estimated that depressive illnesses afflict about 10 percent of adult Americans per year, and depression is one of the ten most likely reasons for someone to consult a family physician. Depression is two to three times more frequent in females than males, and depressive symptoms are more likely in the elderly, with an increase in symptoms occurring after involutional changes of the gonads (menopause, male climacteric). There is also a genetic connection, and depressive disorders and suicide tend to cluster in families; first-degree relatives of depressed patients are up to three times more likely to develop depression than others.[68]

Various names or classifications have been used to describe depressive disorders; however, clinicians are currently using the classification published by the American Psychiatric Association. Several diagnoses of affective disorders are possible, but the pure depressive syndromes are termed *major depression* (severe) and *dysthymia* (mild to moderate) *disorder*. Major depression is often successfully treated with psychotherapy, electroshock, or prescription drugs. Drugs may produce serious side effects but can produce dramatic relief from depressive symptoms.

Dysthymia disorder is a chronic disturbance involving depressed mood and at least two other symptoms, such as appetite or sleep disturbance, fatigue, low self-esteem, hopelessness, poor concentration, and indecisiveness with a history of depressed mood for at least two years.[69] Prescription drug therapy, particularly with the newer selective serotonin reuptake inhibitors such as Prozac, is often used to treat dysthymia disorder. Patient compliance, however, is sometimes a problem because unpleasant transient side effects (headache, gastrointestinal upset, and nervousness) occur in the early phase (one to four weeks) of treatment, and the maximum therapeutic effect may not appear until eight weeks after initiation of therapy. In addition, some drugs have sexual dysfunction side effects that include decreased libido, decreased genital sensitivity, delayed orgasm, and inability to experience orgasm. St. John's wort is an herbal medicine that is effective for treating mild depressive states without the unpleasant side effects associated with prescription drugs.

Herbal remedy for depression

St. John's wort

St. John's wort (SJW) consists of the leaves and flowering tops of *Hypericum perforatum* L. (family Clusiaceae syn. Hypericaceae). The plant is a many-stemmed herbaceous perennial with numerous bright yellow flowers. It is native to Europe and Asia and was brought to the northeastern United States by European colonists. Commercial supplies of the plant are harvested from naturalized plants in the Pacific Northwest and the eastern United States.[70]

In Germany, high-strength preparations of St. John's wort are the leading treatment for mild to moderate depression. Some German physicians prescribe it twenty times more often than Prozac, which is the leading antidepressant used in the United States.

St. John's wort contains numerous compounds with biological activity; however, at this time the active antidepressant principles have not been identified. Most of the research associated with the plant has focused on the pharmacological properties of naphthodianthrones found in the herb in concentrations of up to 0.3 percent. The major naphthodianthrone found is hypericin, but pseudohypericin, isohypericin, protohypericin, and emodin anthrone are also present. In addition, flavonoids such as quercetin, quercitrin, amentoflavone (13, II8-biapigenin), hyperin (hyperoside), and rutin (rutoside) are present in concentrations varying from approximately 7 percent in the leaves and stems up to 11 percent in the flowers. Other constituents include a small amount of an essential oil (0.35 percent) composed of mostly monoterpenes and sesquiterpenes. The final class of compounds is composed of prenylated derivatives of phloroglucinol—namely, hyperforin and adhyperforin.[70]

Among the nine groups of SJW constituents regarded as potential active principles responsible for the plant's antidepressant activity,[71] the most prominent have been the naphthodianthrones, phloroglucinols, and flavonoids. Activity profiles of different SJW preparations in rodent behavioral models—mainly the forced swimming (FST) and tail suspension tests—lent support to the view that the antidepressant activity of

SJW was due to a variety of compounds acting by different mechanisms; flavonoids were prominent among them, with lesser contribution from hyperforin.[72,73] Notably, an extract virtually devoid of both hyperforin and hypericin that was enriched in flavonoids was found to exert significant activity, comparable to hyperforin- and hypericin-containing extracts. This was consistent with the results of a clinical study with the proprietary low-hyperforin (<1 percent)-containing formulation Ze 117 (Max Zeller Söhne AG, Romanshorn, Switzerland), a 50 percent ethanol extract standardized to contain 0.2 percent hypericins.[74]

An interesting observation was made while investigating the activity of a series of ethanolic and methanolic SJW extracts in the FST with rats.[75] All extracts but one showed strong activity; the inactive methanolic extract was found by HPLC analysis to have a much-reduced level of the diglycoside rutin. Addition of rutin to the inactive extract to produce a concentration in the normal range resulted in a pharmacological effect comparable to that of the other original active extracts. The action of rutin appears to be synergistic because the flavonoid itself does not possess antidepressant activity.

In determining the mechanism of action of antidepressive targets such as St. John's wort, it is important to bear in mind that the etiology of depressive disorders is too complex to be explained by a single biologic theory. Patients with major depression, however, have symptoms that reflect changes in brain amine neurotransmitters, particularly norepinephrine, serotonin, or dopamine. Most of the effective antidepressant drugs increase the availability of these neurotransmitters at the synapse; their primary action is either amine reuptake blockade or inhibition of monoamine oxidase (MAO), which catalyzes the degradation of the amine neurotransmitters.[68] In the latter case, early studies on St. John's wort demonstrated in vitro inhibition by hypericin of both types A and B monoamine oxidase.[76] These early findings have not been confirmed, however, and the current thinking is that MAO inhibition as an explanation for the antidepressant activity of St. John's wort can be ruled out.[77]

An intriguing feature held in common by antidepressant drugs including St. John's wort, when they are used in therapy, is a lag phase of up to eight weeks before the full therapeutic effect becomes manifest. For prescription drugs, this lag phase has been postulated to be due to long-term adaptations within the central nervous system that correct imbalances in the brain amine neurotransmitter receptors; this is then translated into a therapeutic effect. Recent animal studies also support this hypothesis for St. John's wort. In one study it was found that after prolonged drug administration in rats, the number of serotonin receptors (both 5-HT_{1A} and 5-HT_{2A}) were significantly increased by 50 percent compared to controls, suggesting an up-regulation.[78] Another study in rats demonstrated a significant up-regulation of 5-HT_2 receptors in the frontal cortex of the brain and a down-regulation of β-adrenergic receptors. In comparison, the tricyclic antidepressant imipramine, also employed in the study, led to a down-regulation of β-adrenergic receptor and 5-HT$_2$-receptor density in the frontal cortex.[79] Although the antidepressant mechanism of St. John's wort is not fully understood, these preliminary findings indicate that it is involved in the same biochemical systems that are relevant in explaining the pharmacological properties of prescription antidepressant drugs.

A meta-analysis of twenty-three randomized trials in a total of 1,757 outpatients with mild or moderate depression found St. John's wort extracts, after two to four weeks, superior to placebo and as effective as standard prescription antidepressants.[80] In most of the studies, however, the diagnosis of depression was not well established, the placebo response rate was lower than expected, the dosage of comparison antidepressants was low, there was a lack of control over patient compliance, and the studies were mostly short term, with the majority lasting no longer than six weeks. A later systematic review and meta-analysis of twenty-one randomized controlled trials of SJW extracts judged SJW superior to placebo, with efficacy comparable to that of standard tricyclic pharmaceuticals. However, the majority of equivalence studies were not of size sufficiently large to allow detection of differences in efficacy less than 20 percent, rendering the matter of equivalence to standard antidepressants still unresolved.[81]

Evaluation of data from sixteen controlled trials of SJW extracts by a scientist at the U.S. National Institutes of Mental Health found them to be superior to placebo against depression, with the number of responders (50 percent) more than twice that of the placebo groups (22 percent).[82] However, the general lack of application of acceptable diagnostic criteria rendered it difficult to determine whether clinically significant antidepressant effects are produced in patients suffering not only from depressive symptoms but also from a depressive disorder. The author of this review noted that the majority of studies lasted for only four weeks and that in only four of the sixteen studies did the number of subjects in the treatment arms exceed fifty. Furthermore, no response to placebo was observed in two studies, whereas 30–60 percent response to placebo normally occurs in clinical trials of antidepressants.

In a review comparing the results of twenty-five controlled studies in a total of 1,592 patients, it was concluded that SJW extract exhibited significant efficacy in the treatment of mild to moderate depression.[83] Comparative studies revealed roughly equivalent responder rates for SJW extracts and synthetic antidepressants, including amitriptyline, bromazepam, desipramine, diazepam, imipramine, and maprotiline. Side effects from SJW were generally less severe and more tolerable. Three studies compared SJW extract to imipramine (one study compared it to placebo as well) for six to eight weeks. The Hamilton Depression (HAMD) Scale showed decreases in both groups, indicating that SJW extract was equivalent to imipramine and is tolerated better by patients for the treatment of mild to moderate depression.

In 2001 and 2002, the results of two randomized controlled trials of SJW, published in the *Journal of the American Medical Association,* sparked intense debate about the plant's effectiveness in major depression[84] or depressive disorder[85]; both studies employed a standardized extract provided by Lichtwer Pharma GmbH, Berlin, Germany. In the former study, a multicenter trial of an SJW extract, two hundred patients (mean age 42.4 years with an average history of depression for more than two years) were diagnosed according to the HAMD scale and DSM-IV (*Diagnostic and Statistical Manual of Mental Disorders,* American Psychiatric Association, 1994). At baseline, participants had a score of at least 20 on the HAMD scale. After a one-week placebo run-in phase during which the most susceptible placebo responders were eliminated, patients were treated with 300 mg t.i.d. of either placebo or SJW for four weeks or more, and 400 mg t.i.d. if improvement was not noted after four weeks. No significant difference was observed in primary outcome measures between SJW and placebo; the only significant difference was a greater incidence of reported headaches (41 percent) in SJW-treated patients, as compared with those on placebo (25 percent).

It has been noted that, considering the reported beneficial effect of high-dose (1,800 mg/day) SJW in major depression, perhaps the dose employed by these researchers (900–1,200 mg/day) was insufficient.[86] It has also been suggested that the patients in this study were too severely and chronically depressed to respond to any medication.[87] In addition, an unusually low placebo response (18.6 percent) was observed. However, according to one definition of remission, a significantly greater percentage of patients responded to SJW (14.3 percent) than to placebo (4.9 percent).

The later trial, conducted in twelve academic and community psychiatric research clinics in the United States, tested the efficacy and safety of a proprietary *H. perforatum* extract against placebo and the SSRI antidepressant sertraline over eight weeks in 340 adult outpatients scoring at least 20 on the HAMD scale.[85] Based on clinical response, the daily dose of SJW ranged from 900 to 1,500 mg and that of sertraline from 50 to 100 mg. The main outcome measures were change in the HAMD total score and the rates of full response determined by the HAMD and Clinical Global Impressions (CGI) scores. The study concluded that efficacy of SJW in moderately severe depression was not supported—with no mention of the observed ineffectiveness of sertraline. The authors of this study speculated that the negative results may have been due to "low assay sensitivity."

As with the earlier trial,[84] critics of this study have suggested that the data imply that the particular group of patients tested had relatively long-term chronic depression and that many of them were already treatment resistant, perhaps explaining the apparent lack of response to sertraline, a well-established pharmaceutical antidepressant. It has also been pointed out that the HAMD entry score for investigational antidepressants is routinely 17 on the 17-item HAMD scale, whereas this study required an entry score of at least 20,[88] indicating severely depressed subjects.

Clearly, more definitive, longer-term studies are required, and the U.S. National Institutes of Health's National Center for Complementary and Alternative Medicine, the National Institute of Mental Health, and the Office of Dietary Supplements are collaborating to fund research to determine the potential benefits and risks of St. John's wort for the treatment of depression.

Even though the active antidepressant constituents are not known, St. John's wort preparations are generally standardized on the basis of hypericin content. The majority of the studies in depression have used an 80 percent methanol extract of St. John's wort standardized to contain 0.3 percent hypericin at a dosage of 300 mg three times daily. Early research focused on hypericin as the chief active antidepressant principle, and it was believed to inhibit MAO A and B. However, subsequent studies revealed inhibitory action of SJW, which was unlikely to account for significant antidepressant activity.[89] If a tea is used, it should be freshly prepared from 2 teaspoonfuls (2–4 g) of finely chopped herb per 150 mL (approx. 2/3 cup) of boiling water; steep ten minutes and strain.[90]

A major advantage of St. John's wort is the minimal incidence of adverse effects compared with other antidepressants. In clinical trials, fewer than 2 percent of patients stopped taking the herb because of adverse effects, and in a four-week open study of 3,250 patients, the most commonly noted side effects were gastrointestinal symptoms (0.6 percent), allergic reactions (0.5 percent), and fatigue (0.4 percent).[91] Hypericin is known to be phototoxic in grazing animals (mainly white or light-coated), resulting in dermatitis of the skin and inflammation of the mucous membranes on exposure to direct sunlight (hypericism).[92]

Because of its potential as a photosensitizing agent, even though this reaction has not been observed in humans with normal doses, it is suggested that therapeutic ultraviolet treatment be avoided when using St. John's wort, and caution regarding exposure to bright sunlight is advised, particularly for fair-skinned persons or those with known photosensitivities. AIDS patients administered 30–40 mg of intravenous hypericin—for its antiretroviral effect—developed phototoxic reactions (facial pain and erythema) after direct exposure to sunlight; this dose was judged equivalent to total hypericin/pseudohypericin in fifty to seventy tablets of the highest dose SJW product.[93] In addition, slight in vitro uterotonic activity from St. John's wort has been reported in animals, suggesting that its use be avoided during pregnancy.[94]

Like most herbs, St. John's wort is valued in folk medicine for a number of conditions unrelated to its principal antidepressant effects. The fresh flowers are crushed and macerated in olive oil, which, after several weeks of standing in the sun, acquires a reddish color. This so-called red oil may be taken internally just like the tea, but it is more commonly applied locally to relieve inflammation and promote healing.

In recent studies, both hypericin and the closely related pseudohypericin have been shown to exhibit antiviral effects in mice infected experimentally with two murine leukemia retroviruses.[95] In addition, both compounds exert effects against a wide spectrum of other viruses, including influenza, herpes simplex types 1 and 2, sindbis, polio, hepatitis C, and murine cytomegalovirus.[94] This, of course, raises the possibility that these compounds from St. John's wort might prove useful in the treatment of AIDS- and HIV-infected patients. Additional studies are certainly warranted.[96]

Drug interactions. St. John's wort affects the bioavailabilty of a substantial percentage of coadministered drugs by influencing their metabolism and absorption. On the metabolic side, the cytochrome P450 (CYP 450) family of isozymes is implicated, particularly CYP 3A4, which is the most abundant hepatic enzyme, claimed to oxidize more than half of all medications subject to oxidative metabolism. CYP 450 enzymes are concentrated in the liver and intestinal mucosa but are also found in the kidneys, lungs, skin, and other tissues. SJW is a potent inducer of CYP 3A within physiologically relevant concentrations—apparently more so in the intestine than in the liver, according to studies with rats. It has been estimated that SJW can effect almost a doubling of CYP 3A4 activity and a corresponding reduction of drug levels, apparently by activation of the pregnane X receptor.

In addition, the adenosine triphosphate (ATP) transporter P-glycoprotein (Pgp) has been identified as a significant contributor to reduction of bioavailability of drugs affected by SJW. An inducible membrane transport protein, Pgp is a barrier to xenobiotic accumulation and a determinant of oral bioavailability of many drugs—notably, anticoagulants, anti-graft-rejection drugs (such as cyclosporine), protease inhibitors (such as the indinaviri used to treat HIV-infected patients), digoxin, and oral contraceptives.[97] Significant reduction of plasma levels of these drugs recommends abstention from SJW during their administration. Although hyperforin and, to a somewhat lesser degree, quercetin[98] are responsible for enzyme and Pgp induction, the low-hyperforin formulation Ze 117 has shown efficacy in mild to moderate depression,[4] while lacking interaction potential with either CYP 3A4 or Pgp.[99]

Pain (general)

Pain is an unpleasant sensory experience associated with actual or potential tissue damage. It may be acute or chronic in character; both types are customarily treated by the administration of drugs known as analgesics. The analgesics used to treat pain are often classified into two categories: the narcotics that bind to opioid receptors in the central and peripheral nervous systems that are used to treat severe pain (sharp or stabbing) and the non-narcotics, such as aspirin, acetaminophen, and the nonsteroidal anti-inflammatory drugs (NSAIDs), which lack an affinity for such receptors. The non-narcotic analgesics are effective for mild to moderate pain (dull or throbbing); for patients with severe pain, these drugs potentiate the effect of narcotic analgesics.[100] Although the most effective herbal analgesics (e.g., morphine and codeine) fall into the first category, they will not be discussed here because they are not available without a prescription.

The non-narcotic analgesics such as aspirin and NSAIDs act by blocking the action of cyclo-oxygenase, which prevents the formation of prostaglandins. When tissues are damaged, a number of pain-mediating substances (e.g., bradykinin and histamine), which excite peripheral nociceptors and send pain signals to the CNS, are released. Prostaglandins sensitize the primary afferent nociceptors to the action of these pain mediators; consequently, inhibition of prostaglandin formation will result in a decrease in pain intensity.

Many of the non-narcotic analgesics used to treat pain are chemically related to salicin—the active constituent of willow bark. The importance of willow bark in the drug development of NSAIDs was discussed in Chapter 1 and illustrated in Figure 1.1.

Herbal remedies for general pain

Willow bark

Use of the bark of *Salix alba* L. and related *Salix* species for their analgesic and anti-inflammatory properties will be discussed in detail in Chapter 9, "Arthritic and Musculoskeletal Disorders." It is sufficient to say here that the active principle, salicin, occurs in the herb in such small amounts as to render its use as a painkiller impractical. Depending upon the salicin content in the bark, between three and twenty-one cups of willow bark tea would have to be consumed to obtain a single average dose. In view of the high tannin content of the bark, this is impractical. A monograph on willow bark was published by the American Herbal Pharmacopoeia in 1999.[101]

Capsicum

Variously known as red pepper, cayenne pepper, and chili pepper, the herb consists of the dried ripe fruit of *Capsicum frutescens* L., *C. annuum* L., and a large number of varieties and hybrids of these members of the family Solanaceae. The medicinal value of capsicum is directly related to its pungency. This varies greatly according to the specific variety involved and its content of capsaicin. Normally, this phenolic derivative occurs in the fruit in concentrations of about 0.02 percent. Its presence accounts for the long-standing folk usage of capsicum as a counterirritant in medicine and a condiment in the culinary arts.[102]

During the past decade, creams containing low concentrations (0.025–0.075 percent) of capsaicin have been found to be effective applied locally to the skin in the treatment of intractable pain such as that associated with herpes zoster (shingles), postmastectomy and postamputation neuroma (phantom pain syndrome), diabetic neuropathy, and even cluster headache (a 0.025 percent cream swabbed intranasally twice daily). The compound causes a depletion of substance P, a neuropeptide that mediates the transmission of pain impulses from the peripheral nerves to the spinal cord. Thus, even if the condition causing the pain continues to be present, no perception of it reaches the brain.

Initial application of capsaicin to the skin produces irritation and hyperesthesia due to release of substance P from peripheral sensory C-type nerve fibers. These fibers are associated with slow, prolonged cutaneous pain transmission. With repeated capsaicin application, substance P is eventually depleted from the nerve fiber, leading to prolonged insensitivity to pain stimuli; consequently, effective use of the cream requires topical application four or five times daily for a period of at least four weeks. Users must be especially careful to wash their hands thoroughly after each application and to avoid touching the eyes or mucous membranes after applying the product.[103] A phytomedicine containing 0.075 percent capsaicin in a cream base has been approved by the FDA for OTC sale.

Headache

Headache, a condition experienced by about 15 percent of the population on a weekly basis, usually is a symptom of an underlying disorder. Almost all headaches are vascular in character or are produced by tension (muscle contraction) or a combination of the two. A small percentage results from underlying intracranial, systemic, or psychologic conditions. Migraine is a type of throbbing vascular headache that affects about 10 percent of Americans and is about three times more prevalent in females than males. The fact that more than half the sufferers of migraines have close relatives who also have them suggests that the tendency may be genetically transmitted. Migraine headaches are generally more severe than tension headaches. A variant vascular headache is the cluster headache, which recurs episodically at intervals of months to years. Its victims are primarily male.

The precise pathophysiologic mechanism that results in migraine headache is poorly understood. Initially, a period of vasoconstriction in the brain may be initiated by some internal or external stimulus such as emotions, stress, glaring lights, noise, smells, or changes in the internal clock. The vasoconstriction causes a reduction in cerebral blood flow that may induce periods of ischemia resulting in neurologic dysfunction. Cerebral vasodilation and neurogenic inflammation follow the ischemic event, resulting in a stimulation of pain receptors.

The biogenic amine serotonin appears to be critical in the vascular changes resulting in migraine. Altering serotonin activity is an important approach in the pharmacotherapy of migraine, and antimigraine drugs are thought to interact predominantly with 5-HT$_1$ and 5-HT$_2$ classes of receptors.[104]

Antimigraine herbs

Treatment of vascular headache involves not only relief of the intense pain of an acute attack but also the prevention (prophylaxis) of additional attacks. During the past decade, research has demonstrated the prophylactic value in the following herbs in treating migraine.

Feverfew

Valued since the time of Dioscorides (a.d. 78) as a febrifuge (antipyretic), the leaves of *Tanacetum parthenium* (L.) Schultz Bip. syn. *Chrysanthemum parthenium* (L.) Bernh. have now been shown to be useful in reducing the frequency and severity of migraine as well as the discomfort of the frequently associated nausea and vomiting. A daily dose of 50–250 mg of dried leaf is recommended for prophylaxis.[105]

Seven randomized double-blind, placebo-controlled clinical trials (RCTs) have been conducted with feverfew monopreparations; five were reported to be successful, three employing dried powdered feverfew leaf.[106–108] Two recent German studies used the same proprietary CO_2 supercritical fluid extract (SFE).[109,110] The first six of these trials have been systematically reviewed.[111] Of the two failed trials, one cannot be properly evaluated because it is available only as an abstract of a poster presentation, with no information as to the particular character of the feverfew preparation or details about outcome measures.[112] The other trial, conducted in the Netherlands, involved a 90 percent alcohol extract of feverfew leaf amply charged (>4 percent) with the then-putative antimigraine principle of feverfew, parthenolide,[113] the germacranolide sesquiterpene lactone, then thought to exert its influence by engaging in Michael addition with free sulfhydryl groups in platelets, thereby inhibiting release of serotonin.[114] The imposition by the Canadian regulatory authority of the criterion of a minimum of 0.2 percent parthenolide for acceptance of the claim as a migraine prophylactic was meant simply to ensure *identity* with the clinically proven sesquiterpene lactone chemotype of feverfew. The failed Dutch trial, contrasted with the later successful Israeli trial[108] with feverfew leaf containing 0.2 percent parthenolide, obviously discredits parthenolide as an appreciable contributor to feverfew's antimigraine activity.

The failure of the Dutch trial with an ethanolic extract of feverfew was likely due to loss of the true, yet unidentified, active principle(s), probably due to degradation during the protracted preparation and processing. The leaf material was macerated at ambient temperature for twelve days with occasional stirring, filtered, and then digested for seven more days! It should also be noted that the two recent trials[109,110] with a feverfew leaf CO_2-SFE product (MIG-99, Schaper & Brümmer GmbH, Salzgitter, Germany) demonstrated primary efficacy in only a small subset of patients who had experienced at least four migraine attacks during the four-week baseline period preceding the trials. The scientific community awaits the clinical trial of an undegraded hydroalcoholic extract of feverfew for migraine prevention.

No serious adverse effects have been associated with consumption of feverfew. In a survey of roughly 270 people who had been using feverfew for two to four years (more than 70 percent of whom claimed less frequent or less painful headaches, or both) conducted by the City of London Migraine Clinic, 18 percent reported adverse events. The most troublesome was the formation in the mouth of recurrent, so-called aphthous, ulcers experienced by 11.3 percent and severe enough in 7 percent as to prompt discontinuation of therapy; 6.5 percent of respondents reported experiencing indigestion. About 40 percent of users, however, attributed pleasant side effects to feverfew, noting, among other experiences: a sense of well-being, relief of the symptoms of coexisting arthritis, and tranquilizing effects such as more restful sleep and reduced muscular tension.[115]

Aphtous ulceration is apparently a systemic effect that resolves within a week or so of discontinuation of the treatment but returns on rechallenge.[116] Interestingly, the clinical trial conducted at the University of Nottingham[107] reported that more patients in the placebo group ($n = 16$) reported mouth ulceration than in the verum group ($n = 10$). Also interesting is the observation that the nonsteroidal anti-inflammatory drugs (NSAIDs) increasingly used for migraine prophylaxis and arthritis also cause recurrent aphthous ulceration; the effect occurs particularly with drugs that produce substantial inhibition of the cyclooxygenase pathway.[117] It has been claimed that this mouth ulceration can be alleviated by treatment with tincture of myrrh (*Commiphora* sp.).[118] The bitter-tasting feverfew leaves sometimes induce a more generalized inflammation of the oral mucosa and tongue, with attendant swelling of the lips and loss of taste. The soreness is likely caused by direct contact with leaves during chewing and is probably due to interaction with sesquiterpene lactones known to cause contact dermatitis.[119]

A "postfeverfew syndrome" has been identified in about 10 percent of long-term feverfew users who stopped taking the herb; about one-tenth experienced moderate to severe aches, pains, and stiffness in joints and muscles, along with CNS symptoms of anxiety and poor sleep.[106] It has been speculated that such sleep disturbances may be due to withdrawal of melatonin, present in appreciable quantities in the leaf (2.45 µ/g in fresh; 2.19 µ/g in dried).[67]

In view of feverfew's traditional reputation as an emmenagogue to induce uterine contraction in full-term women and its ability to cause abortion in cattle, it would seem prudent to forego its use during pregnancy or lactation. It should be noted, however, that when feverfew has been used to promote menstruation, it was taken in much higher doses than currently employed for treating migraine and arthritis. Finally, little is known of the effects of feverfew on migraine and arthritis in pregnancy.[116] There are no reports of feverfew causing teratogenic side effects and no data on its safety during lactation. A study involving thirty females who had been consuming feverfew for more than eleven months revealed no differences in the frequency of chromosomal aberrations or the frequency of sister chromatid exchanges compared with a matched set of nonusers.[120]

Feverfew is also contraindicated in persons with recognized hypersensitivity to other members of the Asteraceae (Compositae) because cross-reactivity is common among plants in this family.[121] No substantial information on feverfew use by children is available.

Butterbur

Extracts of both leaves and root/rhizome of *Petasites hybridus* (L.) Gaertn, Meyer & Scherb. (Asteraceae/Compositae) have been clinically tested as migraine prophylactic as well as treatment for seasonal allergic rhinitis. The Swiss company Zeller AG manufactures an extract of butterbur leaves (Ze339), standardized to contain 25–35 percent petasins (sesquiterpene esters). Marketed as Tesalin™, a root extract, Petadolex™ is produced by Weber and Weber in Petaforce™ and is available from Bioforce Ltd., Irvine, United Kingdom. These manufacturers claim that the hepatotoxic and potentially carcinogenic pyrrolizidine alkaloids (PAs) of butterbur have been removed.[122] However, preparations from the rhizome of *P. hybridus* were removed from the market in Switzerland in early 2004.[123]

Nonetheless, Petadolex was submitted to clinical trials for prophylaxis of migraine.[124–126] Both migraine studies noted a significant reduction in frequency of migraine attacks, as well as excellent tolerability, recommending butterbur as a effective migraine prophylactic.

A report published in 2003 assessed the safety of Petadolex.[127] Results from acute, subchronic, and chronic animal toxicity studies, as well as safety data from clinical trials, post-marketing surveillance studies, and pharmacovigilance, were evaluated and discussed. The patented butterbur root extract was judged safe for treatment in humans. Thus far, only four cases of a reversible cholestatic hepatitis have been probably associated with long-term administration of butterbur (incidence of 1:175,000).[128]

Ginger

One case history from Denmark in 1990 reported the beneficial effect of ginger in aborting the effects of migraine headache: a forty-two-year-old female migraineur given 1.5–2.0 g of powdered ginger daily at the onset of a migraine attack experienced marked reduction in the frequency and severity of migraine headache.[129] Recently, a proprietary product (Gelstat Migraine) has been claimed effective as a sublingually administered feverfew and ginger compound for treatment of acute migraine during the mild pain phase.[130]

Cannabis

Cannabis sativa L., Cannabaceae, was once a widely accepted medical treatment for the prevention and relief of migraine headache, listed in the United States Pharmacopeia from 1860 to 1941.[131] Savitex® (GW Pharmaceuticals, Salisbury, UK) is a cannabis-based blend of whole plant extracts that delivers approximately equal amounts of Δ-9-tetrahydrocannabinol (THC) and cannabidiol (CBD). THC has analgesic activity in both nociceptive and neuropathic pain, and both THC and CBD exert anti-inflammatory effects. Savitex has been approved for several medicinal applications in the United Kingdom and is also available by prescription in Canada for symptomatic relief of neuropathic pain in multiple sclerosis (MS). Clinical trials conducted in the United Kingdom support the benefit of the cannabis-based medicine in MS[132] and rheumatoid arthritis.[133]

Caffeine-containing beverages

These CNS-stimulant plant products are most commonly employed simply to overcome drowsiness; in addition, they are used therapeutically as adjuncts in the treatment of headache. The acute consumption of caffeine in conjunction with OTC analgesics such as aspirin or acetaminophen increases their activity by as much as 40 percent, depending on the specific type of pain involved. This beneficial effect is apparently due to the ability of caffeine to cause constriction of the cerebral blood vessels and, possibly, to facilitate the absorption of other drugs. The enhancement is short-lived, diminishing greatly on repeated (chronic) coadministration.

The principal caffeine beverages are prepared from

- coffee—dried ripe seed of *Coffea arabica* L. (family Rubiaceae);
- tea—leaves and leaf buds of *Camellia sinensis* (L.) O. Kuntze (family Theaceae);
- kola (cola)—dried cotyledons of *Cola nitida* (Vent.) Schott & Endl. and other *Cola* spp. (family Sterculiaceae);

- cacao (cocoa)—roasted seeds of *Theobroma cacao* L. (family Sterculiaceae);
- guarana—crushed seed of *Paullinia cupana* Kunth. ex H. B. K. (family Sapindaceae); and
- maté—dried leaves of *Ilex paraguariensis* St.-Hil. (family Aquifoliaceae).

Guarana has the highest caffeine content of all caffeine-containing plants—up to 5 percent.[134]

Research from Austria's Medical University of Innsbruck has shown that caffeine boosts short-term memory and task orientation. Using MRI scans, the Innsbruck researchers discovered that subjects who consumed approximately 100 mg of caffeine—about the amount in two cups of coffee—had higher activity levels in the frontal lobe, an important memory center, and the anterior cingulum, which controls attention span. The effect typically lasts for about twenty minutes.[135]

Other research on coffee suggests its benefits go beyond mental acuity. Work cited by CoffeeScience.org points to beneficial antioxidants in coffee—four times those found in green tea—as the onset of type II diabetes.

Depending on its exact mode of preparation, one cup of beverage prepared from these herbs will contain amounts of caffeine ranging from about 10 mg for cocoa to some 30 mg for tea to approximately 100 mg for coffee. So-called cola-flavored carbonated beverages are prepared using only minute amounts of cola. They do contain about 50 mg per bottle of caffeine, which is added as a "flavor." In the United States, caffeine is also added to a variety of food products, including baked goods, frozen dairy desserts, gelatin puddings, soft candy, and the like.[136]

Caffeine-containing beverages are used in combination with salicylates for the treatment of headache or with ergot alkaloids (ergotamine) for migraine. The customary dose is a quantity equivalent to 100–200 mg of caffeine.[137]

It should also be noted that caffeine and the other naturally occurring xanthine derivatives—theobromine and theophylline—possess a diuretic action, but it is relatively weak and of short duration. Neither caffeine nor the various beverages containing it are extensively used for this purpose because tachyphylaxis develops rapidly on continued administration—a frequent situation with such drinks—and diuretic effectiveness is greatly diminished.

Toothache

Tooth pain usually results from a pathologic condition of the dentin, the pulp, or the supporting periodontum. The manifestations most amenable to temporary treatment with phytomedicinal agents are those resulting from caries or fractures. Either of these events results in an exposure of the dentin that, when subjected to heat, cold, or pressure, stimulates the numerous free nerve endings in the pulp, resulting in pain. If the pulp itself is exposed, the ache is continuous; if the pulp becomes infected, the pain will also be severe.[138]

The proper recourse for any toothache is timely treatment by a qualified dentist. This might consist of filling, root canal therapy, extraction, periodontal treatment, or the like. However, if professional care is not available, the pain may be allayed by the temporary use of appropriate herbal remedies. Such action must be viewed as only an emergency measure because prolonged use of self-selected remedies may exacerbate the underlying condition causing the pain.

Herbal remedies for toothache

Clove oil

Probably the best-known herbal product used to obtain transient relief from toothache is clove oil. This is obtained by steam distillation from clove (cloves), the dried flower bud of *Syzygium aromaticum* (L.) Merr. & T.M. Perry of the family Myrtaceae. For many years, the correct scientific name of the plant was considered to be either *Eugenia caryophyllata* Thunb. or *Caryophyllus aromaticus* L. These names are still encountered in the older herbal literature. All three designations refer to the same species. Clove of good quality yields about 15–25 percent of a volatile oil that has both local analgesic and antiseptic properties. These are due to a number of phenolic substances contained in the oil, the principal one of which is eugenol, constituting 85 percent of the total.

Although clove oil is sufficiently irritating to preclude general internal usage, it has long been employed as a local analgesic or obtundent for the relief of toothache. Eugenol, like other phenols, acts on contact to depress cutaneous sensory receptors involved in pain perception. Additional analgesic activity results from pronounced inhibition of prostaglandin and leukotriene biosynthesis resulting from blockages of the cyclooxygenase and lipoxygenase metabolic pathways.[139] In practice, a pledget of cotton is dipped in the undiluted oil and applied to the surface of the aching tooth and surrounding tissue or, if possible, inserted directly into the cavity, where it will alleviate the pain for several hours. The oil is also used in mouthwashes, in concentrations of 1–5 percent, for its antiseptic effects.

Various forms of clove (actually listed as cloves), including the volatile oil, appear as safe food additives on the GRAS (generally recognized as safe) list of the Food and Drug Administration. Because they can cause damage to viable pulp and soft tissue when applied indiscriminately to an aching tooth, the American Dental Association has accepted clove oil, or its constituent eugenol, for professional use only, not for nonprescription use. In Germany, Commission E has approved the use of clove oil as a local anesthetic and antiseptic.[140]

Prickly ash bark

The barks of two species of *Zanthoxylum* are used more or less interchangeably as toothache remedies in the United States. Both *Z. americanum* Mill., the northern prickly ash, and *Z. clava-herculis* L., the southern prickly ash, are sometimes referred to as toothache trees. These members of the family Rutaceae both yield prickly barks that, when chewed, produce a tingling sensation in the mouth and are also effective remedies for toothache. For all practical purposes, the barks of the two species may be considered as a single herb. It was valued by the American Indians, who chewed the bark and then packed the masticated quid around the ailing tooth to relieve the pain.[141]

Although prickly ash bark, under the title Xanthoxylum, held official status in the USP and then the NF from 1820 to 1947 and was employed as a diaphoretic and antirheumatic, the principles responsible for its local anesthetic effect appear not to have been investigated scientifically. Older sources simply attribute the bark's pungency when it is chewed to its resin content.[142] More recent studies on a related West African species that has similar properties, *Z. zanthoxyloides* (Lam.) Watson, associate at least some of its anti-inflammatory action with the presence of fagaramide, an aromatic acid amide. Fagaramide apparently acts as an inhibitor of prostaglandin synthesis and might account for some of the analgesic effects of the bark as well, although it is approximately twenty times less potent than indomethacin.[143] It is not known if the American *Zanthoxylum* species contain fagaramide.

Prickly ash bark is used for toothache today in the same manner as it was employed long ago by Native Americans. A small amount of the bark is chewed, and the moist mass is packed around the painful tooth as an emergency method of relieving pain. Adverse effects have not been reported from short-term use.

Sexual impotence

The inability of a male to attain or maintain penile erection sufficient to complete intercourse is termed erectile dysfunction or, more commonly, impotence. Psychogenic factors such as sexual anxieties, guilt, fear, and feelings of inadequacy are responsible for 50–60 percent of erectile dysfunction. The remaining such cases are caused by organic factors.[144]

Herbal remedies for sexual impotence

Yohimbe

An herb that holds some promise in the treatment of erectile dysfunction is yohimbe. Consisting of the bark of the West African tree *Corynanthe johimbe* K. Schum. syn. *Pausinystalia yohimbe* (K. Schum.) Pierre of the family Rubiaceae, yohimbe contains about 6 percent of a mixture of alkaloids, the principal one of which is yohimbine. Both yohimbe and yohimbine have long enjoyed considerable reputations as aphrodisiacs.

Mechanistically, yohimbine is primarily an α_2-adrenergic antagonist. Its peripheral effect is to increase cholinergic and decrease adrenergic activity. In male sexual performance, erection is linked to cholinergic activity, which results in increased penile blood inflow, decreased outflow, or both, causing erectile stimulation.[145]

A review of clinical studies to determine the value of yohimbine in treating impotency shows mixed results. Assessment of the efficacy of yohimbine is complicated by relatively high rates of positive response in placebo-treated subjects. In addition, some studies have described a two- to three-week latency period before full therapeutic effect. In general, the therapeutic gain achieved with yohimbine in clinical trials can only be described as modest.[146] For example, in one trial, some 46 percent of a group of patients being treated for psychogenic impotence reported a positive response to the drug, and 43 percent of patients with organic causes of impotence reported improvement in erectile function.[147]

Additional difficulties with using yohimbine are associated with pharmacokinetic problems. In a study to examine tolerability and pharmacodynamics of single doses of yohimbine, significant inter-subject variability in the pharmacokinetic parameters was observed. Plasma levels of yohimbine varied among individuals, probably due to differences in the rate of first-pass hepatic metabolism of the drug. Furthermore, in participants fed a high-fat meal, plasma levels were decreased by 30 percent, demonstrating that a fasting stomach may be best for oral administration.[148]

Tablets containing 5.4 mg of yohimbine hydrochloride are currently being marketed in the United States as a prescription drug. It is prescribed to treat male erectile dysfunction; however, the U.S. Food and Drug Administration has not approved yohimbine for this use.

The German Commission E notes that proof of the activity of yohimbe and yohimbine is insufficient and that the risk–benefit ratio unacceptably high; the commission does not recommend their therapeutic use. Side effects include agitation, tremors, insomnia, anxiety, hypertension, tachycardia, nausea, and vomiting.[149] In spite of this ruling, a large number of aphrodisiac preparations containing yohimbine continue to be marketed in Germany.

In the United States, yohimbe products were purchased at retail health food outlets by field investigators of the FDA. Upon capillary gas chromatography analysis of twenty-six products, it was found that nine products contained no yohimbine and eighty-nine contained only trace amounts (0.1–1 ppm). Products purporting to be only yohimbe bark would be expected to contain 7,000–10,000 ppm yohimbine.[150] These studies clearly indicate that the majority of products in the marketplace are of inferior quality.

Ginkgo

The reports that prolonged oral administration of *Ginkgo biloba* extract (GBE; see Chapter 6) may benefit persons suffering from erectile dysfunction are highly preliminary.[151] Additional data are required to support the possible use of GBE for this purpose.

References

1. *Drug evaluations annual 1993.* 1992. Chicago: American Medical Association, 213–217.
2. *Diagnostic and statistical manual of mental disorders* (DSM-IV-R), 4th ed., 393–444. Washington, D.C.: American Psychiatric Association.
3. Chrisman, M. L., and D. M. Jermain. 1996. Sleep aid and stimulant products. In *Handbook of nonprescription drugs,* 11th ed., ed. T. R. Covington, 179–192. Washington D.C.: American Pharmaceutical Association.
4. Trevor, A. J., and W. L. Way. 1995. Sedative-hypnotic drugs. In *Basic and clinical pharmacology,* 6th ed., ed. B. G. Katzung, 333–349. Norwalk, CT: Appleton & Lange.
5. Hocking, G. M. 1997. *A dictionary of natural products,* 836–837. Medford, NJ: Plexus Publishing, Inc.
6. Morazzoni, P., and E. Bombardelli. *Fitoterapia* 66:99–112.
7. Bisset, N. G., ed. 1994. *Herbal drugs and phytopharmaceuticals,* English ed. (*Teedrogen,* M. Wichtl, ed.), 513–516. Boca Raton, FL: CRC Press.
8. Upton, R., ed. 1999. Valerian root, *Valeriana officinalis.* Analytical, quality control and therapeutic monograph, American Herbal Pharmacopoeia, Santa Cruz, CA, p. 25.
9. Krieglstein, J., and D. Grusla, D. 1988. *Deutsche Apotheker Zeitung* 128:2041–2046.
10. Hiller, K.-O., and G. Zetler. 1996. *Phytotherapy Research* 10:145–151.
11. Newall, C. A., L. A. Anderson, and J. D. Phillipson. 1996. *Herbal medicines: A guide for health-care professionals,* 260–262. London: The Pharmaceutical Press.
12. McKenna, D. J., K. Jones, and K. Hughes. 2002. *Botanical medicines. The desk reference for major herbal supplements,* 2nd ed., 1007–1037. New York: The Haworth Press, Inc.
13. *Bundesanzeiger* (Cologne, Germany): May 15, 1985; March 6, 1990.
14. Willey, L. B., S. P. Mady, D. J. Cobaugh, and P. M. Wax. 1995. *Veterinary and Human Toxicology* 37:364–365.
15. Singh, Y. N., and M. Blumenthal. 1997. *HerbalGram* 39:33–55.

16. Bone, K. 1993/1994. *British Journal of Phytotherapy* 3:147–153.
17. Norton, S. A., and P. Ruze. 1994. *Journal of the American Academy of Dermatology* 31:89–97.
18. *Review of Natural Products:* November 1996.
19. Hänsel, R. 1996. *Zeitschrift für Phytotherapie* 17:180–195.
20. Davies, L. P., C. A. Drew, P. Duffield, G. A. R. Johnston, and D. D. Jamieson. 1992. *Pharmacology and Toxicology* 71:120–126.
21. Gleitz, J., A. Beile, and T. Peters. 1995. *Neuropharmacology* 34:1133–1138.
22. Murray, M. T. 1995. *The healing power of herbs,* 2nd ed., 210–219. Rocklin, CA: Prima Publishing.
23. Volz, H.-P., and M. Kieser. 1997. *Pharmacopsychiatry* 30:1–5.
24. *Bundesanzeiger* (Cologne, Germany): June 1, 1990.
25. Almeida, J. C., and E. W. Grimsley. 1996. *Annals of Internal Medicine* 135:940–941.
26. Anke, J., and I. Ramzan. 2004. *Planta Medica* 70:193–196.
27. Zou, I., M. R. Harkey, and G. K. Henderson. 2005. *Planta Medica* 71 (2): 142–146.
28. Bauer, R., B. Kopp, and A. Nahrstedt. 2003. *Planta Medica* 69:971–972.
29. Ross, M. S. F., and L. A. Anderson. 1986. *International Journal of Crude Drug Research* 24:1–6.
30. *Lawrence Review of Natural Products:* May 1989.
31. Speroni, E., and A. Minghetti. 1988. *Planta Medica* 54:488–491.
32. *Federal Register* 43 (114): 25578 (1978).
33. *Bundesanzeiger* (Cologne, Germany): November 30, 1985; March 6, 1990.
34. Hänsel, R., and H. H. Wagener. 1992. *Arzneimittel-Forschung* 17:79–81.
35. Hölzl, J. 1992. *Zeitschrift für Phytotherapie* 13:155–161.
36. Mannering, G. J., and J. A. Shoeman. 1996. *Xenobiotica* 26:487–493.
37. Deepak, M., G. K. Sangli, P. C. Arun, and A. Amit. 2005. *Phytochemical Analysis* 16:24–29.
38. Singh, R. H., and L. Singh. 1980. *Journal of Research in Ayurveda and Siddha* 1:133–148.
39. Bhattacharya, S. K, and S. Ghosal. 1998. *Phytomedicine* 5:77–82.
40. Rai, D., G. Bhatia, G. Palit, R. Pal, S. Singh, and H. K. Singh. 2003. *Pharmacology, Biochemistry and Behavior* 75:823–830.
41. Anonymous. 2004. *Alternative Medicine Review* 9 (1): 79–85.
42. Stough, C., J. Lloyd, J. Clarke, L. A. Downey, C. W. Hutchison, T. Rodgers, and P. J. Nathan. 2001. *Psychopharmacology* 156 (4): 481–484.
43. Nathan, P. J., J. Clarke, J. Lloyd, C. W. Hutchison, L. Downey, and C. Stough. 2001. *Human Psychopharmacology* 16:345–351.
44. Singh, H. K., and B. N. Dhawan. 1997. *Indian Journal of Pharmacology* 29 (5): 5359–5365.
45. Awang, D. V. C. 1998. *Canadian Pharmaceutical Journal* 131 (7): 42–46.
46. *Herba Centellae.* 1999. *WHO monographs on selected medicinal plants,* vol. 1, 77–85. Geneva: World Health Organization.
47. Kartnig, T. 1998. *Journal of Herbs, Spices & Medicinal Plants* 3:146–173.
48. Gupta, S. S., S. Shinde, V. L. Iyengar, and S. Shastry. *Indian Journal of Pharmacology* 12:64.
49. Diwan, P. V., I. Karwande, and A. K. Singh. 1991. *Fitoterapia* 62 (3): 253–257.
50. Sarma, D. N. K., R. L. Khosa, J. P. N. Chansauria, and M. Sahai. 1996. *Phytotherapy Research* 10 (2): 181–183.

51. Bradwejn, J., Y. Zhou, D. Koszycki, and J. Shlik. 2000. *Journal of Clinical Psychopharmacology* 20 (6): 680–684.
52. Veerendra Kumar, M. H., and Y. K. Gupta. 2002. *Journal of Ethnopharmacology* 79:253–260.
53. Appa Rao, M. V. R., K. Srinivasan, and T. Koteswara Rao. 1977. *Indian Journal of Psychiatry* 19 (4):54–49.
54. Lentner, C., ed. 1981. *Geigy scientific tables,* vol. 1, 235. Basle: Ciba–Geigy.
55. Hartmann, E. 1978. *The sleeping pill,* 162–181. New Haven, CT: Yale University Press.
56. Reynolds, J. E. F., ed. 1996. *Martindale: The extra pharmacopoeia,* 31st ed., 336–337. London: The Royal Pharmaceutical Society of Great Britain.
57. Mayeno, A. N., and G. J. Gleich. 1994. *Trends in Biotechnology* 12:346–352.
58. Simat, T., B. Van Wickern, K. Eulitz, and H. Steinhart. 1996. *Journal of Chromatography B* 685:41–51.
59. Nightingale, S. L. 1992. *Journal of the American Medical Association* 268:1828.
60. Webb, S. M., and M. Puig-Domingo. 1995. *Clinical Endocrinology* 42:221–234.
61. Garfinkel, D., M. Laudon, D. Nof, and N. Zisapel. 1995. *Lancet* 346:541–544.
62. Zhdanova, I. V., R. J. Wurtman, H. J. Lynch, J. R. Ives, A. B. Dollins, C. Morabito, J. K. Matheson, and D. L. Schomer. 1995. *Clinical Pharmacology and Therapeutics* 57:552–558.
63. Generali, J. A. 1996. *Drug Newsletter* 15:3–5.
64. Van den Heuvel, C. J., K. J. Reid, and D. Dawson. 1997. *Physiology and Behavior* 61:795–802.
65. Crismon, M. L., and D. M. Jermain. 1996. Sleep aid and stimulant products. In *Handbook of nonprescription drugs,* 11th ed., ed. T. R. Covington, 179–192. Washington, D.C.: American Pharmaceutical Association.
66. *Lawrence Review of Natural Products:* January 1996.
67. Murch, S. J., C. B. Simmons, and P. K. Saxena. 1997. *Lancet* 350:1598–1599.
68. Wells, B. G., L. A. Mandos, and P. E. Hayes. 1997. Depressive disorders. In *Pharmacotherapy: A pathophysiologic approach,* 3rd ed., ed. J. T. DiPiro, R. L. Talbert, G. C. Yee, G. R. Matzke, B. G. Wells, and L. M. Posey, 1395–1417. Stamford, CT: Appleton & Lange.
69. *Diagnostic and statistical manual of mental disorders* (DSM-IV-R), 4th ed., 339–349. 1994. Washington, D.C.: American Psychiatric Association.
70. Upton, R., ed. 1997. St. John's wort, *Hypericum perforatum,* American herbal pharmacopoeia and therapeutic compendium, 1–32. Santa Clara, CA, published in *HerbalGram* 40.
71. Rey, J. M., and G. Walter. 1998. *Medical Journal of Australia* 169:583–586.
72. Bombardelli, E., and P. Morazzoni. 1995. *Fitoterapia* 66:43–68.
73. Butterweck, V., V. Christoffel, A. Nahrstedt, F. Petercit, B. Spengter, and H. Winterhoof. 2003. *Life Sciences* 73:627–639.
74. Käufeler, R., B. Meier, and A. Brattström. 2001. *Pharmacopsychiatry* 34 (suppl. 1): 549–550.
75. Nöldner, M., and K. Schötz. *Planta Medica* 68:577–580.
76. Suzuki, O., Y. Katsumata, M. Oya, S. Bladt, and H. Wagner. 1984. *Planta Medica* 50:272–274.
77. Bladt, S., and H. Wagner. 1994. *Journal of Geriatric Psychiatry and Neurology* 7 (suppl. 1): S57–S59.
78. Teufel-Mayer, R., and J. Gleitz. 1997. *Pharmacopsychiatry* 30 (suppl.): 113–116.

79. Müller, W. E., M. Rolli, C. Schäfer, and U. Hafner. 1997. *Pharmacopsychiatry* 30 (suppl.): 102–107.
80. Linde, K., G. Ramirez, C. D. Mulrow, A. Pauls, W. Weidenhammer, and D. Melchart. 1996. *British Medical Journal* 313:253–258.
81. Whiskey, E., U. Werneke, and D. Taylor. 2001. *International Clinical Psychopharmacology* 16 (5): 239–252.
82. Vitiello, B. 1999. *Journal of Pharmacy and Pharmacology* 51: 513–517.
83. Harrer, G., and V. Schulz. *Journal of Geriatric Psychiatry and Neurology* 7 (suppl.) S6–S8.
84. Shelton, R. C., M. B. Keller, A. Gelenberg, D. L. Dunner, R. Hirschfeld, M. E. Thase, et al. 2001. *Journal of the American Medical Association* 285 (15): 1978–1986.
85. Hypericum Depression Trial Study Group (Davidson, J. R. T. et al.). 2002. *Journal of the American Medical Association* 287:1807–1814.
86. Vorbach, E. U., E. U. Arnoldt, and W.-D. Hübner. 1997. *Pharmacopsychiatry* 30 (suppl.): 81–85.
87. Cott, J. M. 2001. *Journal of the American Medical Association* (letter) 286 (1): 42.
88. Jonas, W., D. Wheatley, G. I. Spielman, A. Völp, K. Linder, D. Melchart, C. D. Mulrow, M. Berner, and J. Cott. *Journal of the American Medical Association* (letters) 288:466–449.
89. Cott, J. M. 1997. *Pharmacopsychiatry* 30 (suppl. 2): 108–112.
90. Bisset, N. G., ed. 1994. *Herbal drugs and phytopharmaceuticals,* English ed. (*Teedrogen,* M. Wichtl, ed.), 273–275. Boca Raton, FL: CRC Press.
91. Woelk, H., G. Burkard, and J. Grunwald. 1994. *Journal of Geriatric Psychiatry and Neurology* 7 (suppl. 1): S34–S38.
92. Roth, L. 1990. *Hypericum–Hypericin: Botanik, Inhaltsstoffe, Wirkung,* ecomed, Landsberg/Lech, Germany, 135–138.
93. Gulick, R. M., V. McAuliffe, J. Holden-Wiltse, C. Crumpacker, L. Liebes, D. S. Stein, et al. 1999. *Annals of Internal Medicine* 130:510–514.
94. *Review of Natural Products:* November 1997.
95. Muruelo, D., G. Lavie, and D. Lavie. 1988. *Proceedings of the National Academy of Science, U.S.A.* 85:5230–5234.
96. James, J. S. 1990. *AIDS Treatment News* 117:3.
97. Cott, J. 2002. *Alternative Therapies in Women's Health* 4 (8): 60–64.
98. Weber, C. C., S. Kressmann, G. Firicker, and W. E. Müller. 2004. *Pharmcopsychiatry* 37 (6): 292–298.
99. Brattström, A. 2002. *Deutsche Apotheker Zeitung* 142 (30): 3695–3699.
100. Baumann, T. J. 1997. Pain management. In *Pharmacotherapy: A pathophysiologic approach,* ed. J. T. DiPiro, R. L. Talbert, G. C. Yee, G. R. Matzke, B. G. Wells, and L. M. Posey, 1259–1278. Stamford, CT: Appleton & Lange.
101. Upton, R., ed. 1999. Willow bark, *Salix* spp. In *American herbal pharmacopoeia and therapeutic compendium,* Santa Clara, CA, 18.
102. Robbers, J. E., M. K. Speedie, and V. E. Tyler. 1996. *Pharmacognosy and pharmacobiotechnology,* 134–135. Baltimore, MD: Williams & Wilkins.
103. Rumsford, J. A., and D. P. West. 1991. DICP. *Annals of Pharmacotherapy* 25:381–387.
104. Beckett, B. E. 1997. Headache disorders. In *Pharmacotherapy: A pathophysiologic approach,* ed. J. T. DiPiro, R. L. Talbert, G. C. Yee, G. R. Matzke, B. G. Wells, and L. M. Posey, 1279–1291. Stamford, CT: Appleton & Lange.
105. Awang, D. V. C. 1998. *Integrative Medicine* 1 (1): 11–13.

106. Johnson, E. S., N. P. Kadam, D. M. Hylands, and P. J. Hylands. 1985. *British Medical Journal* 291:569–573.
107. Murphy, J. J., S. Heptinstall, and J. R. A. Mitchell. 1998. *Lancet* ii:189–192.
108. Palevitch, D., G. Earon, and R. Carasso. 1997. *Phytotherapy Research* 11:508–511.
109. Pfaffenrath, V., H. C. Diener, M. Fisher, M. Friede, and H. H. Heinnecke-von Zepelin. 2002. *Cephalalgia* 22:523–532.
110. Diener, H. C., V. Pfaffenrath, J. Schnitker, M. Friede, and H. H. Hennecker-von Zepelin. 2005. *Cephalalgia* 25:1031–1041.
111. Pittler, M. H., and E. Ernst. 2004. *The Cochrane Library issue 1.* Chichester, England: John Wiley & Sons, Ltd.
112. Kuritzky, A., Y. Elhacham, Z. Yerushalmi, and R. Hering. 1994. *Neurology* 44 (suppl. 2): A201 (abstract 293 P).
113. de Weerdt, C. J., H. P. R. Bootsma, and H. Hendriks. 1996. *Phytomedicine* 3:225–230.
114. Heptinstall, S., and D. V. C. Awang. 1998. In *Phytomedicines of Europe: Chemistry and biological activity,* ed. L. D. Lawson and R. Bauer, 158–175. ACS Symposium Series 691. Washington, D.C.: American Chemical Society.
115. Johnson, E. S. 1983. *MIMS Magazine* (May 15): 32–34.
116. Johnson, S. 1984. *Feverfew—A traditional remedy for migraine and arthritis,* 118. London: Sheldon Press.
117. Fenton, D. A., E. R. Young, and J. D. Wilkinson. *British Medical Journal* 286:1062 (letter).
118. Gardiner, A. 1996. *Medicinal herbs and essential oils,* 24. London: Promotional Reprint Co. Ltd.
119. Hausen, B. M., E. Busker, and R. Carle. 1984. *Planta Medica* 34:229–235.
120. Anderson, D., P. C. Jenkinson, R. S. Dewdney, S. D. Blowers, E. S. Johnson, and N. P. Kadam. 1988. *Human Toxicology* 7:145–152.
121. Hausen, B. M., and P. E. Osmundesn. 1983. *Acta Dermatoloica Venereologia* 63:308–314.
122. Brattström, A. 2004. Ze 339 *Schwezerische Zeitschrift für Ganzheitsmedizin* 16(4): 158–160.
123. Schlenger, R. 2004. *DeutscheA apotheker Zeitungg* 144 (8): 64–66.
124. Grossmann, M., and H. Schmidramsl. 2000. *International Journal of Clinical Pharmacology and Therapy* 38 (9): 430–435; *Alternative Medical Review* 6 (3): 303–310 (2001).
125. Diener, H. C., V. W. Rahlfs, and U. Danesch. 2004. *European Neurology* 51 (2): 89–97.
126. Lipton, R. B., H. Göbel, K. M. Einhäupl, K. Wilks, and A. Muskop. 2004. *Neurology* 63:2240–2244.
127. Danesch, U., and R. Rittinghausen. 2003. *Headache* 43 (1): 76–68.
128. Kalin, P. 2003. *Forschende Komplementarmedizin und Klassische Naturheilkunde* 10 (suppl. 1): 42–44.
129. Mustafa, T., and K. C. Srivastava. 1990. *Journal of Ethnopharmacology* 29:267–273.
130. Cady, R. K., C. P. Schreiber, M. E. Beach, and C. C. Hart. 2005. *Medical Science Monitor* 11 (9): 165–169.
131. Russo, E. 1998. *Pain* 76:3–8.
132. Rog, D. J., T. J. Nurmikko, T. Friede, and C. A. Young. 2005. *Neurology* 65 (6): 812–819.

133. Blake, D. R., P. Robson, M. Ho, R. W. Jubb, and C. S. McCabe. 2006. *Rheumatology* 45:50–52.
134. Duke, J. A. 1985. *Handbook of medicinal herbs*, 349. Boca Raton, FL: CRC Press.
135. Anonymous. *The National Post* (Canada), December 5, 2005.
136. Tyler, V. E. 1993. *The honest herbal*, 3rd ed., 53–56. Binghamton, NY: Pharmaceutical Products Press.
137. Haas, H. 1991. *Arzneipflanzenkunde*, 44–45. Mannheim, Germany: B. I. Wissenschaftsverlag.
138. Flynn, A. A. 1996. Oral health. In *Handbook of nonprescription drugs*, 11th ed., ed. T. R. Covington, 509–535. Washington, D.C.: American Pharmaceutical Association.
139. Deininger, R. 1991. *Zeitschrift für Phytotherapie* 12:205–212.
140. *Bundesanzeiger* (Cologne, Germany): November 30, 1985.
141. Mizelle, R. 1986. *Encounter with the toothache tree*. Kitty Hawk, NC: Carolina Banks Publishing, 52 pp.
142. Osol, A., and G. E. Farrar, Jr. 1947. *The dispensatory of the United States of America*, 24th ed., 1650–1651. Philadelphia, PA: J. B. Lippincott.
143. Newall, C. A., L. A. Anderson, and J. D. Phillipson. 1996. *Herbal medicines: A guide for health-care professionals*, 219–221. London: The Pharmaceutical Press.
144. Borirakchanyavat, S., and T. F. Lue. 1997. Erectile dysfunction. In *Conn's current therapy*, ed. R. E. Rakel, 694–697. Philadelphia, PA: W. B. Saunders Co.
145. Robbers, J. E., M. K. Speedie, and V. E. Tyler. 1996. *Pharmacognosy and pharmacobiotechnology*, 169. Baltimore, MD: Williams & Wilkins.
146. Riley, A. J. 1994. *British Journal of Clinical Practice* 48:133–136.
147. Reid, K., D. H. C. Surridge, A. Morales, M. Condea, C. Harris, J. Owen, and J. Fenemore. 1987. *Lancet* 2:421–423.
148. Grasing, K., M. G. Sturgill, R. C. Rosen, J. R. Trout, T. J. Thomas, G. D. Kulkarni, P. Maines, J. R. and Seibold. 1996. *Journal of Clinical Pharmacology* 36:814–822.
149. *Bundesanzeiger* (Cologne, Germany): August 14, 1987; February 1, 1990.
150. Betz, J. M., K. D. White, and A. H. der Marderosian. 1995. *Journal of AOAC International* 78:1189–1194.
151. *Sex Over 40* 10 (4): 6–7 (1991).

chapter eight

Endocrine and metabolic problems

Hormones in endocrine and metabolic disorders

Hormones are chemical messengers that transmit information from one cell to another to coordinate growth, development, and reproduction in the body. They are synthesized in a limited number of tissues and secreted into the circulation, where they are transferred to a targeted tissue. Here they bind with receptor proteins that possess a recognition site, which binds hormones with high specificity, and an activity site, which translates the information received into a biochemical message.

Two classes of hormones operate via two types of receptors. Peptide hormones and amino acid–derived hormones act via receptors located in the cell membrane and often activate second messengers that amplify and distribute the molecular information. Steroid hormones and thyroid hormones interact with intracellular receptors that bind to DNA recognition sites to regulate transcription of target genes. They change the concentration of cell proteins, primarily enzymes, which results in a change in the metabolic activity of the cell.

The secretion and production of hormones in the body are generally under close regulatory control. As illustrated in Figure 8.1, they involve a feedback control of production and a hormonal cascade that begins with signals in the central nervous system followed by hormone secretion by the hypothalamus, pituitary, and end target gland. Hormones produced by the end target gland (e.g., sex hormones and corticosteroids) feed back on the hypothalamic–pituitary system to regulate their own rates of synthesis. When the endogenous pool of a particular hormone increases to a certain level in the plasma, the hypothalamus or the pituitary gland stops production of the specific releasing factor or tropic hormone, which in turn stops production of the particular hormone by the target gland.

The endogenous pool of hormone is also dependent on the rates of metabolism and excretion of the hormone. Because of feedback control of hormone secretion, changes in rates of hormone degradation generally do not cause endocrine pathology, provided the feedback control mechanisms that regulate synthesis are intact. For example, in severe liver disease, the degradation of corticosteroids by the liver is impaired; however, the endogenous pool level of corticosteroids does not increase above the

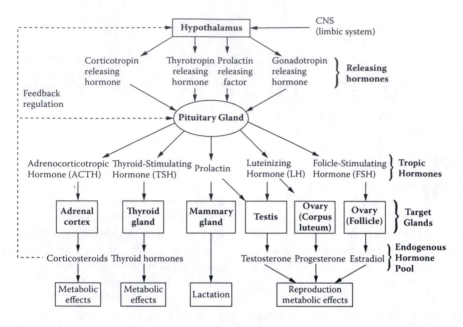

Figure 8.1 Hormonal cascade and feedback regulation in the production of hormones.

normal range because secretion of ACTH by the pituitary is inhibited (Figure 8.1).[1]

The most prevalent endocrine disorders result from hormone deficiencies. In addition to the involutional changes of the gonads associated with the aging process, a variety of disease states impairs or destroys the ability of endocrine glands to synthesize and secrete hormones. These include defects in gland development, genetic defects in biosynthetic enzymes, immune mediated destruction, neoplasia, infections, nutritional deficits, and vascular insufficiency. As a consequence, endocrine therapy often involves hormone replacement such as the use of insulin to treat diabetes or estrogen to alleviate the symptoms of menopause.[1]

In general, the ability of a hormone to react with its particular receptor is very specific and is associated with rigid chemical structure requirements for the hormone molecule in order for it to fit the receptor and produce a biological effect. An analogy often used to describe this phenomenon is the precise configuration required for a key to fit and open a lock. For this reason, hormone replacement therapy usually involves natural hormones or chemicals very closely related in structure.

Unfortunately, many people are under the misconception that plants produce steroid hormones. This erroneous idea probably stems from the fact that plants such as the wild yam (*Dioscorea*) and soya bean produce steroid compounds that have no hormonal activity but, rather, serve as starting materials in the chemical synthesis of steroid hormones. The case of progesterone and wild yam is illustrative of the extreme foolishness that can result from an incomplete understanding of the biology, chemistry, and production of steroid hormones. Diosgenin, a steroidal sapogenin produced in wild yam, is a starting material for the chemical synthesis of progesterone.[2,3] Diosgenin and wild yam are not known to exert estrogenic activity in humans, yet wild yam herbal products are falsely advertised and sold to treat conditions requiring progesterone replacement.[4]

Estrogenic activity, on the other hand, is unique in that it does not require a strict structural configuration as do the other sex hormones and the corticosteroids. The estrogen receptor has been found to accommodate a diverse array of aromatic structural types that do not contain the steroid nucleus. Medicinal chemists have made available a large number of therapeutically useful nonsteroidal estrogen agonists such as diethylstilbesterol (DES), which is used as a postcoital contraceptive agent, and antagonists such as tamoxifen, used in the treatment of breast cancer. In addition, the constituents of many plants have weak estrogenic activity and are called phytoestrogens. Phytoestrogens are found in more than twenty different classes of phytochemicals, including cooumestans, isoflavones, lignans, resorcyclic acid lactones, including the fungal mycotoxins (zearalenones and their derivatives) produced by molds of *Fusarium* species that commonly infect corn, wheat, barley, sorghum, hay, and even ginseng.[5] Prominent examples of these are the isoflavonoid genistein, found in soy beans and other legumes, and the coumarin derivative coumesterol, found in a number of different legumes.

Considered the most potent of phytoestrogens, coumestrol is actually the collective name for the twenty or so coumestans thus far identified. It is considered to be thirty to one hundred times more active as an estrogen than the isoflavones, about two hundred times less potent than estrone, and almost three thousand times less potent than DES.[6] In vitro studies with these compounds have shown a weak binding to the estrogen receptor—roughly 0.4 and 13 percent, respectively—relative to 100 percent binding affinity for the natural ligand estradiol.[7] Assays with hepatocytes demonstrate that phytoestrogens have potencies ranging between one thousand and two thousand times less than estradiol; however, this low potency does not mean phytoestrogens will be inactive in therapy. Because of their high therapeutic index, much larger doses can be administered.[8]

The revelation by the Women's Health Initiative that hormone replacement therapy (HRT) with equine estrogen, combined with progestin, was associated with a significantly increased risk of developing breast cancer[9] led to 56 percent of women discontinuing use of HRT. That event generated much increased interest in alternative treatments for menopausal symptoms such as hot flashes, anxiety, insomnia, and osteoporosis.

Phytoestrogens appeared a promising alternative to HRT and research into their activity consequently increased.

More than twenty classes of phytochemicals are natural human estrogen receptor (ER) ligands; the five most potent are steroids, polyketides (zearalenones), alkylated flavanones, isoflavones, and phenylbenzofurans. The most researched of these compounds have been the isoflavones, identified as the phytoestrogenic principles of soy and red clover (*Trifolium pretense* L.) and responsible for observed in vitro and in vivo activity suggestive of clinical potential. However, clinical data on phytoestrogens are sparse. In an exhaustive review of the literature through 2003, only two clinical trials were identified concerning evaluation of another promising phytoestrogenic plant, *Humulus lupulus* L. (hops; see discussion in Chapter 7). A recent review of the pharmacognosy of hops emphasizing its estrogenic properties concluded that "hops preparations which contain 8-prenylnaringenin (8-isopentenylnaningenin) or hopein must be considered estrogenic."[10]

It has yet to be determined, however, whether such preparations can have beneficial hormonal activity when consumed orally by humans. No officially recognized standards exist for estrogenic formulations of hops.

An interesting phytoestrogenic controversy regarding conflicting reports of ginseng's estrogenic activity has been explained on the basis of fungal contamination: Root extracts of Asian (*Panax ginseng*) and American (*P. quinquefolius*) ginseng were found to bind to both ERα and ERβ, with two to three times greater affinity for the latter. Subsequent analysis of the extracts revealed significant ER binding attributable to zearalenone, the estrogenic mycotoxin produced by several *Fuarium* fungal species. The ERs showed no binding affinity for ginsenoside Rb1, the major ginsenoside of both ginseng species, or for Rg1, a prominent ginsenoside in Asian ginseng.[5]

Plant constituents may have hormonal activity through mechanisms different from replacement therapy. As can be noted in Figure 8.1, many places in the hormonal cascade can be influenced by the inhibitory activity of a plant constituent. One of these is the endogenous hormone pool. The level of particular hormones in the pool is governed by enzymes that promote either the synthesis or degradation of hormones. Compounds that inhibit these enzymes, thus increasing or decreasing the amount of hormone, would appear to be acting as a hormone or an antihormone.[11]

For example, in the case of the undesirable mineralocorticoid activity of licorice (Chapter 3), the active constituent glycyrrhetinic acid inhibits 11β-hydroxysteroid dehydrogenase, an enzyme that degrades hydrocortisone (a compound with mineralocorticoid activity) to cortisone, which has very little mineralocorticoid activity. Inhibition of this conversion increases the level of hydrocortisone in the endogenous corticosteroid pool, with a subsequent increase in undesirable mineralocorticoid activity. The net level of corticosteroids in the endogenous pool is not increased, so there is no influence on the feedback regulation loop of the hormonal cascade. The general outcome of the use of licorice would lead one to conclude that it contains an active constituent that has intrinsic mineralocorticoid activity, but it actually inhibits a degradative enzyme and indirectly produces a mineralocorticoid effect by upsetting the balance of hydrocortisone to cortisone found in the endogenous hormone pool.

Gynecological disorders

These include a variety of conditions, such as menopausal symptoms, premenstrual syndrome including mastodynia, and dysmenorrhea. All are directly or indirectly related to imbalances or deficiencies in the production of female sex hormones or prostaglandins.

In the female menopause—the period following the complete cessation of menstruation—estrogen production drops to about 10 percent of its premenopausal levels and progesterone production drops to nearly zero. This results in various symptoms, four of which are more or less common: (1) vasomotor disorders (hot flashes), (2) urogenital atrophy, (3) osteoporosis, and (4) psychological disturbances. Administration of conjugated equine estrogens has been a common treatment for menopausal symptoms.[12] However, hormone replacement therapy (HRT) has been associated with a slight but significant increase in the risk of developing breast and endometrial cancer,[13] leading to marked increase in the use of herbal remedies as alternative therapy.[14]

Premenstrual syndrome (PMS) is characterized by certain emotional and physical symptoms as well as behavioral changes that occur during the premenstrual (luteal) phase of the menstrual cycle and then disappear several days after the onset of menstruation. Its etiology is unknown; however, premenstrual symptoms are closely linked to the rise and fall of gonadotropin and ovarian hormones, neurotransmitters, and prostaglandins during the menstrual cycle. Some studies suggest that normal fluctuations in estrogen and progesterone trigger the symptoms by an unknown mechanism. The beginning of premenstrual symptoms parallels the postovulatory rise in progesterone levels, but the highest severity of symptoms occurs five to six days after the peak of progesterone levels. When estradiol levels peak during the preovulatory phase, women feel better and have minimal symptoms. The rate of decline in progesterone and estradiol during the late luteal phase has been proposed as being more important in causing symptoms than absolute basal values. Stress may play a significant role, and some women appear to be predisposed to PMS by various genetic, biological, or psychological factors.[15]

Primary dysmenorrhea—that is, painful menstruation in the absence of pelvic pathology—occurs with considerable frequency, particularly among adolescent females. Pain usually begins with, or just slightly before, the onset of menstrual flow and lasts for periods of up to two days, seldom longer.

It has now been shown that this condition is associated with an increased production and concentration of prostaglandin in the endometrium during the luteal and menstrual phases of the cycle. This results in hypercontractility of uterine muscle during dysmenorrhea. Prostaglandin inhibitors have proven to be very effective in relieving the associated symptoms.[15]

The existence of naturally occurring steroidal estrogens in species other than vertebrates is uncertain; however, it is estimated that more than three hundred plants possess compounds with estrogenic activity. A few of these phytoestrogenic herbs have utility in treating health problems associated with gynecological disorders.

Herbal remedies for gynecological disorders

Black cohosh

The dried rhizome and roots of *Actaea racemosa* L. (syn. *Cimicifuga racemosa* (L.) Nutt.), family Ranunculaceae, is sometimes referred to as black snakeroot or cimicifuga. It has an ancient reputation as a remedy for the treatment of "female complaints," a generic term that probably includes all of the gynecological disorders just discussed. As such, it was one of the ingredients in the famous proprietary medicine, Lydia E. Pinkham's Vegetable Compound.[16]

Hänsel has reviewed the literature supporting claims of estrogen-like activity for extracts of black cohosh.[17] A clinical study of hysterectomized patients with climacteric symptoms by Lehmann-Willenbrock and Riedel showed no significant differences among groups treated with various estrogens and those receiving black cohosh extracts.[18] The beneficial effects were slow to appear, requiring up to four weeks to reach a maximum.

More recently, investigators have shown that an alcoholic extract of black cohosh suppressed hot flashes in menopausal women by reducing the secretion of luteinizing hormone (LH). It also suppressed LH production in ovariectomized rats. Synergistically acting compounds are thought to be responsible for this effect, but they remain unidentified.[19] The drug does contain the isoflavone formononetin, which may be the basis of the estrogenic activity because it is converted in vivo by the gastrointestinal microbial flora to compounds that bind to the estrogen receptor—first to daidzein and then to the more active isoflavan equol. In addition, the steroidal triterpene glycosides actein, 27-deoxyactein, and cimicifugoside have been identified, but their role in the gynecological activity of black cohosh is unknown.[20]

German Commission E has found black cohosh to be effective for the treatment of PMS and dysmenorrhea, as well as for nervous conditions associated with menopause.[21] The herb is normally administered in the form of a 40–60 percent alcoholic extract in a quantity equivalent to 40 mg of drug daily. A decoction prepared from 0.3–2.0 g of the herb may also be employed. Administration of the drug sometimes causes stomach upsets; otherwise, no problems or contraindications have been reported. In view of the fact that no long-term toxicity studies on the use of black cohosh have been carried out, administration of the herb should be limited to a period of no longer than six months.

Chasteberry

The aromatic fruit of *Vitex agnus-castus* L., a small deciduous tree or large shrub of the family Verbenaceae that grows in the Mediterranean region, has long been used in European herbal medicine, but it is little known in the United States. In recent years, however, an extract of the fruit has become available in that country.[22]

Chasteberry is now believed to have dopaminergic properties. Dopamine and dopamine agonists inhibit the secretion of the peptide hormone prolactin by the pituitary gland.[15,23] Exactly what effect this has on PMS, dysmenorrhea, and menopause is not entirely clear, but the symptoms of PMS, including mastodynia as well as amenorrhea and irregular menses, have been associated with elevated blood levels of prolactin. Drugs that reduce prolactin concentrations usually restore the menstrual cycle to normal.[15]

The therapeutic benefit of a single daily dose of a 20-mg commercial aqueous-ethanolic extract of vitex fruits was investigated in a randomized, double-blind, placebo-controlled trial. Fifty-two women with disturbances of the menstrual cycle were chosen for the study. Following a three-month treatment period, women receiving vitex extract had a significant reduction in prolactin release in response to tropic hormone stimulation compared to placebo. In addition, women in the vitex group had a significant reduction in PMS symptoms, while the placebo group did not. No side effects were reported.[24] The nature of the chemical principles in chasteberry responsible for this prolactin-depressant effect has not been established. Preliminary isolation studies utilizing vitex extracts containing water-soluble substances, pressure filtration through molecular sieves, and column chromatography have led to the identification of at least three different dopaminergic compounds that potently inhibit prolactin release from cultured rat pituitary cells.[25]

The German Commission E has recommended the use of chasteberry for a variety of menstrual disturbances and mastodynia.[26] The herb is usually administered in the form of a concentrated alcoholic extract of the fruit; average dose is 20 mg per day. Patients are warned that the herb should not be used during pregnancy or lactation, and the use of the herb does occasionally produce an itchy rash in sensitive consumers.

Evening primrose oil

The small seeds of this native American wildflower, *Oenothera biennis* L. (family Onagraceae) contain about 14 percent of a fixed oil of which 70 percent is *cis*-linoleic acid and 9 percent is *cis*-gamma-linolenic acid (GLA). The latter constituent is a relatively uncommon one that is found in quantity in only a few other plants, such as black currant and borage seeds.[27-29]

Theoretically, GLA can be converted directly to the prostaglandin precursor dihomo-GLA (DGLA). Therefore, administration of the oil containing it might be beneficial to persons unable to metabolize *cis*-linolenic acid to GLA and to produce subsequent intermediates of considerable metabolic significance, including the less inflammatory PGE_1, which can lead to an imbalance in the ratio of inflammatory to noninflammatory prostaglandin compounds.[30] Illnesses thought by some to arise from such metabolic deficiencies and imbalances—and therefore presumably treatable by the administration of evening primrose oil—are numerous. The more significant ones in terms of supporting evidence are PMS and associated mastodynia, as well as atopic eczema. However, even in these conditions, which have been the subject of a number of studies, the results are controversial.

A comprehensive literature review[31] has reported that evening primrose oil had some utility in treating PMS and mastodynia. In the latter case, the rationale was that low levels of PGE_1 may increase the effect of prolactin on breast tissue, causing mastodynia. On the other hand, a recent report describing a literature search of clinical trials of evening primrose oil for the treatment of PMS with a view to performing a meta-analysis found only five trials where randomization was clearly indicated. Inconsistent scoring and response criteria made a meta-analysis inappropriate. The two best-controlled studies failed to show any beneficial effects for evening primrose oil in the management of PMS.[32]

Clinical trials testing the effects of the oil in the treatment of atopic eczema have also provided conflicting results.[31]

In view of these uncertainties regarding the efficacy of evening primrose oil and the relatively high cost of the 500-mg capsules (ca. $0.25 each—minimum dose of four per day), one cannot unreservedly recommend the use of this product. This position is further supported by the lack of toxicity data regarding its long-term use.

GLA-rich seed oils from two other plant sources—black currant oil and borage seed oil—are currently available in the United States.

Black currant oil

Seeds of *Ribes nigrum* L.(family Grossulariaceae), the European black currant, yield a fixed oil containing 14–19 percent GLA. Capsules containing approximately 200 or 400 mg of the product are currently marketed.

Borage seed oil

Obtained from the seeds of *Borago officinalis* L., family Boraginaceae, this oil contains 20–26 percent GLA.[29] It is currently marketed in the form of capsules, each containing 1,300 mg of the oil, equivalent to 300 mg of GLA. Borage seeds have been shown to contain small amounts of pyrrolizidine alkaloids, including the known hepatotoxin amabiline. That alkaloid was not detected in samples of the seed oil down to levels of 5 µg/g. Consumption of 1–2 g of borage seed oil daily could, nevertheless, result in an intake of toxic unsaturated pyrrolizidine alkaloids (UPAs) approaching 10 mg. The German federal health agency now limits internal consumption of such products to not more than 10 µg of UPA daily.[33]

The fact that GLA-rich fatty oils are now available commercially from three different plant sources does not alter the conclusion that the product's efficacy for any condition remains unproven. Likewise, its safety for long-term usage requires additional verification.

Raspberry leaf

Raspberry leaf tea, an infusion prepared from the dried leaves of *Rubus idaeus* L. or *R. strigosus* Michx. of the family Rosaceae, has a considerable reputation as "a traditional remedy for painful and profuse menstruation and for use before and during confinement to make parturition easier and speedier."[34] Because of its astringent properties, it is also used to treat diarrhea, an application previously discussed in Chapter 3. The scientific evidence supporting the effects of raspberry leaf on the uterus is scanty, and clinical evidence is even more so. However, a 1999 study reported a positive influence on labor outcomes.[35]

Beckett and colleagues have carried out the most substantial pharmacological testing to date, using isolated tissues of guinea pigs and frogs.[36] They concluded that aqueous raspberry leaf extracts contain a number of different active constituents, the actions of which are mutually antagonistic, including (1) a smooth muscle stimulant, (2) an anticholinesterase, and (3) a spasmolytic. The authors opined that it would be impossible to predict an overall clinical effect from observation of or studies with animals.

There is also a difference in the effect of the herb on pregnant versus nonpregnant human uterine strips. It was without effect on the latter but promoted contraction of normal human uterine strips at ten to sixteen weeks of pregnancy.[37]

Such findings seriously complicate any attempt at the evaluation of the efficacy of the herb for any of its folkloric uses. Evaluation is further hampered by the absence of any long-term toxicity data, including teratogenicity. This is an especially serious omission in view of the fact that one of the recommended uses of raspberry leaf tea is to alleviate morning sickness.[38] Doses also are relatively large. At least one source recommends preparing a tea by steeping 30 g of leaves in 480 mL of boiling water for thirty minutes and drinking the entire quantity to promote an easy labor.[39]

In view of these factors, it seems best to adhere to the admonition that the consumption of any herbal product of unproven safety and efficacy is especially unwise during pregnancy. Deviation from this rule will certainly produce more harm than good.

Hyperthyroidism

Also referred to as thyrotoxicosis or Graves' disease, hyperthyroidism is a very common endocrine disorder that may arise from a number of different causes. It is characterized by symptoms that may include weakness, weight loss, nervousness, tachycardia, exophthalmos, and goiter.

Herbal remedy for hyperthyroidism

Bugle weed
An herb that has been used to treat the condition is bugle weed. It consists of the leaves and tops collected before flowering of *Lycopus virginicus* L. or *L. europaeus* L. of the family Lamiaceae. Several investigations dealing with the pharmacological activity of bugle weed have demonstrated pronounced antithyrotropic activity in vitro and in vivo. The activity is purportedly due to phenolic plant constituents; the most active is rosmarinic acid.[40] Oral administration of an ethanolic extract of *Lycopus europaeus* to rats caused a decrease of triiodithyronine (T_3) levels, presumably as a consequence of a reduced peripheral thyroxine (T_4) deiodination, as well as a pronounced reduction of T_4 and thyroid stimulating hormone (TSH), indicating a central point of attack of the plant extract.[41] This preliminary scientific evidence suggests that bugle weed may have application in treating hyperthyroidism; however, carefully controlled clinical trials are needed before the use of bugle weed can be recommended. It is also important to emphasize that hyperthyroidism is a complex disease scarcely amenable to self-treatment with nonstandardized medicaments.

Diabetes mellitus

Diabetes is characterized by inappropriate hyperglycemia resulting from a deficiency of insulin, a reduction in its effectiveness, or both. The condition is normally treated by diet, the administration of exogenous insulin, or the use of oral hypoglycemic drugs. The latter function, at least in part, stimulates the β-cells of the pancreas to produce more insulin.

A large number of plants have been shown to exert hypoglycemic effects in small animal studies. However, most of these plants have not been adequately tested in human beings to demonstrate conclusively their safety and utility as a substitute for insulin or for the oral hypoglycemic drugs.[42] Further, none of them is currently marketed in the form of a preparation with standardized activity. This would be an absolute necessity if they were to be used to control hyperglycemia successfully.

During the past decade, the results of preliminary trials assessing the antidiabetic potential of a number of herbal preparations have been published. Forty diabetic Thai women, aged thirty-five to sixty years, were administered 1 tablespoon of *Aloe vera* gel juice (80 percent) twice a day for at least two weeks. Blood sugar and triglyceride levels in the treated group were significantly reduced compared to the control group on "a carminative mixture" (unspecified); cholesterol levels were not affected.[43] A study involving thirty men and thirty women with type 2 diabetes, aged 52.2 ± 6.3 years, randomly assigned subjects to six groups. Groups 1, 2, and 3 consumed 1, 3, or 6 g of ground cinnamon (*Cinnamomun cassia* J. Presl, family Lauraceae; 1 g = 2 × 500-mg capsules) daily, respectively, for forty days; this was followed by a twenty-day washout period. Groups 4, 5, and 6 were given corresponding numbers of capsules containing wheat flour as placebo. Mean fasting serum glucose levels were reduced after forty days by all three levels of cinnamon intake (18–29 percent), as were levels of triglycerides (23–30 percent), LDL cholesterol (7–27 percent), and total cholesterol (12–26 percent); no significant changes were noted in the placebo group.[44]

The effect of fenugreek (*Trigonella foenum-graecum* L., family Fabaceae) seeds on glycemic control and insulin resistance in mild type 2 diabetes was assessed in a double-blind, placebo-controlled trial with twenty-five newly diagnosed patients. Group 1 (*n* = 12) received 1 g/day of hydroalcoholic extract of fenugreek seeds; group 2 (*n* = 13) received usual care (dietary control, exercise) and placebo capsules for two months. Fasting blood glucose and two-hour postglucose blood glucose were not significantly different in the two groups, but area under the curve (AUC) of blood glucose as well as insulin was significantly lower ($p < 0.001$). HOMA model-derived insulin resistance showed a decrease in percent β-cell secretion in group 4 as compared to group 2 (86.3 ± 32 vs. 70.1 ± 52) and increase in percent insulin sensitivity (112.9 ± 67 vs. 92.2 ± 57) ($p < 0.05$). Also, serum triglycerides decreased and HDL cholesterol increased significantly in group 1 as compared to group 2 ($p < 0.05$).[45]

The long-term effect of supplementation with tablets of spirulina (*Spirulina platensis*), a blue-green algae, on blood sugar levels, serum lipid profiles, and glycated protein levels in non-insulin-dependent diabetes mellitus (NIDDM) (type 2 DM) was assessed in fifteen NIDDM patients (seven males, eight females); the control group consisted of seven NIDDM patients (four males, three female). After two months, patients in the experimental group, who received 2 g of spirulina daily, experienced a 34 percent decrease in glycated serum proteins and a significant (p = 40.01) decrease (27 percent) in fasting blood sugar. Triglyceride levels were decreased by 22 percent, free fatty acids by 34 percent, and total cholesterol by 11 percent.[46]

Other herbal remedies for diabetes mellitus

Ginsengs

Clinical evidence exists for both American ginseng (*Panax quinquefolius* L.) and Asian ginseng (*P. ginseng* C. A. Meyer) exerting hypoglycemic effects. Recent randomized controlled studies conducted in Canada with American ginseng in groups of ten and twelve subjects found that 1–3 g of powdered whole root, taken any time up to two hours before a glucose meal, led to a roughly 20 percent reduction in blood sugar level for patients suffering from type 2 diabetes.[47,48] A similar effect was noted in nondiabetics, except that the American ginseng had to be consumed at least forty minutes before the sugar meal; blood glucose level was not affected in the absence of glucose challenge,[49,50] thus posing no risk of hypoglycemia in nondiabetics.

Ginsenosides, the characteristic triterpene saponins of *Panax* species, are widely regarded as chiefly responsible for many of the roots' activities, including hypoglycemic effects. However, the individual ginsenosides have been shown to have different, sometimes opposing, activities.[51] In a randomized, single-blind design, twelve normal subjects (six females and six males), aged 31 ± 3 years, were administered 6 g of a batch of American ginseng root, which was different from that used in earlier successful antidiabetic trials, or placebo forty minutes before a 75-mg oral glucose tolerance test. Repeated measures of analysis of variance demonstrated no significant effect of the ginseng on incremental plasma glucose or insulin or their AUC indices of insulin sensitivity.[52] These results suggest that changes in ginsenoside levels and profile play a role in the hypoglycemic effects; the authors of the study note particularly marked decrements in total ginsenosides and changes in the key protopanaxadiol to protopanaxatriol ratios, Rb_1:Rg_1 and Rb_2:Rc.

Widely publicized studies with mice, reported in two publications in 2002, extolled the effectiveness of a *Panax ginseng* berry extract as a treatment for diabetes and obesity; drastic reduction in blood glucose levels ("completely normalized") and loss of 10–15 percent body weight in twelve days, as well as a 30 percent reduction of cholesterol levels, were observed in treated diabetic mice, which ate 15 percent less food and were 35 percent more active than untreated ones.[53,54] The authors of the studies, conducted at the University of Chicago, attributed the antidiabetic effects—but not the antiobesity effect—to content of ginsenoside Re in the intraperitoneally administered berry extract, which contained five to seven times more of the ginsenoside than root extract of Asian ginseng at a level comparable to the content of American ginseng root.

An apparent Asian ginseng root extract preparation, of unspecified composition, administered to thirty-six NIDDM patients has also been reported to reduce fasting blood glucose levels.[55]

Further research is needed to elucidate properly the mechanisms by which antihyperglycemic botanicals exert their antidiabetic effects. Also needed are well-designed, larger clinical trials of longer duration to firmly establish safety and efficacy.

Apart from the ultimate aim of elucidation of mechanisms of action, there is an insistent need to develop bases for standardization that tie the composition of herbs to their efficacy. The lack of a reliable basis for ginseng led Vuksan and Sievenpiper[56] to use an acute postprandial clinical screening model to select an efficacious ginseng batch, dose, and time of administration. Upon reviewing the published evidence in support of a broad variety of herbs used in diabetes, these two authors have concluded that "compelling evidence" of efficacy from randomized controlled clinical trials exists for only one herb other than American ginseng—namely, ivy gourd [*Coccinia grandis* (L.) J. Voigt, syn. *C. indica* Wight & Arn., Cucurbitaceae].

References

1. Wilson, J. D. 1998. Approach to the patient with endocrine and metabolic disorders. In *Harrison's principles of internal medicine*, 14th ed., ed. A. S. Fauci, E. Braunwald, K. J. Isselbacher, J. D. Wilson, J. B. Martin, D. L. Kasper, S. L. Hauser, and D. L. Longo, 1965–1972. New York: McGraw–Hill.
2. Robbers, J. E., M. K. Speedie, and V. E. Tyler. 1996. *Pharmacognosy and pharmacobiotechnology*, 123–124. Baltmore, MD: Williams & Wilkins.
3. Huxtable, R. J., ed. 1996. *Proceedings of the Western Pharmacological Society* 39:1–6.
4. Hudson, T. 1996. *Townsend Letter for Doctors and Patients* July: 125–127.
5. Gray, S. L., B. R. Lackey, P. L. Tate, M. B. Riley, and N. D. Camper. 2004. *Experimental Biological Medicine* 229:560–568.
6. Beckham, N. 1998. *Australian Journal of Medical Herbalism* 7:11–16.
7. Miksicek, R. J. 1994. *Journal of Steroid Biochemistry and Molecular Biology* 49:153–160.
8. Pelissero, C., G. Flouriot, J. L. Foucher, B. Bennetau, J. Dunogues, F. LeGac, and J. P. Sumpter. 1993. *Journal of Steroid Biochemistry and Molecular Biology* 44:263–272.

9. Possouw, J. E., G. L. Anderson, R. L. Prentice, A. Z. Croix, C. Kooperberg, M. L. Stefanick, et al. 2002. *Journal of the American Medical Association* 288:321–333.
10. Chadwick, L. R., G. F. Pauli, and N. R. Farnsworth. 2006. *Phytomedicine* 13:119–131.
11. Baker, M. E. *Proceedings of the Society for Experimental Biology and Medicine* 208:131–138.
12. Mullins, P. M., M. C. Pugh, and A. O. Moore. 1997. Hormone replacement therapy. In *Pharmacotherapy: A pathophysiologic approach*, 3rd ed., ed. J. T. DiPiro, R. L. Talbert, G. C. Yee, G. R. T. Matzke, B. G. Wells, and L. M. Posey, 1635–1646. Stamford, CT: Appleton & Lange.
13. Stampfer, M. J., W. C. Willett, D. J. Hunter, and J. E. Manson. 1992. *Cancer Causes & Control* 3:33–39.
14. Murkies, A. L., G. Wilcox, and S. R. Davis. 1998. *Journal of Clinical Endocrinological Metabolism* 83:297–303.
15. Fankhauser, M. P. 1997. Premenstrual syndrome. In *Pharmacotherapy: A pathophysiologic approach*, 3rd ed., ed. J. T. DiPiro, R. L. Talbert, G. C. Yee, G. R. Matzke, B. G. Wells, and L. M. Posey, 1621–1633. Stamford, CT: Appleton & Lange.
16. Burton, J. 1949. *Lydia Pinkham is her name*, 107. New York: Farrar, Straus.
17. Hänsel, R. 1991. *Phytopharmaka*, 2nd ed., 223–230. Berlin: Springer–Verlag.
18. Lehmann-Willenbrock, E., and H. H. Riedel. 1988. *Zentralblatt für Gynäkologie* 110:611–618.
19. Düker, E.-M., L. Kopanski, H. Jarry, and W. Wuttke. 1991. *Planta Medica* 57:420–424.
20. Newall, C. A., L. A. Anderson, and J. D. Phillipson. 1996. *Herbal medicines: A guide for health-care professionals*, 80–81. London: The Pharmaceutical Press.
21. *Bundesanzeiger* (Cologne, Germany): January 5, 1989.
22. Blumenthal, M. 1989. *Health Foods Business* 35 (9): 18, 104–105.
23. Winterhoff, H., C. Gorkow, and B. Behr. 1991. *Zeitschrift für Phytotherapie* 12:175–179.
24. Miulewicz, A., E. Gejdel, H. Sworen, K. Sienkiewicz, H. Jedrzejak, J. Teucher, and H. Schmitz. 1993. *Arzneimittel-Forschung* 43:752–756.
25. Wuttke, W., C. Gorkow, and H. Jarry. 1995. Dopaminergic compounds in *Vitex agnus-castus*. In *Phytopharmaka in Forschung und Klinischer Anwendung*, ed. D. Loew and N. Rietbrock, 81–91. Darmstadt, Germany: Steinkopf.
26. *Bundesanzeiger* (Cologne, Germany): May 15, 1985.
27. Briggs, C. J. 1986. *Canadian Pharmaceutical Journal* 119:248–254.
28. Barber, A. J. 1988. *Pharmaceutical Journal* 240:723–725.
29. Awang, D. V. C. 1990. *Canadian Pharmaceutical Journal* 123:121–126.
30. Newall, C. A., L. A. Anderson, and J. D. Phillipson. 1996. *Herbal medicines: A guide for health-care professionals*, 110–113. London: The Pharmaceutical Press.
31. *Review of Natural Products:* August 1997.
32. Budeiri, D., A. LiWan Po, and J. C. Dornan. 1996. *Controlled Clinical Trials* 17:60–68.
33. Commission E, German Federal Health Agency. 1992.
34. *Martindale—The extra pharmacopoeia*, 31st ed., 1748. London: The Royal Pharmaceutical Society, 1996.
35. Parsons, M., M. Simpson, and T. Ponton. 1999. *Australian College of Midwives International Journal* 12:20–25.

36. Beckett, A. H., F. W. Belthle, K. R. Fell, and M. F. Lockett. 1954. *Journal of Pharmacy and Pharmacology* 6:785–796.
37. Bamford, D. S., R. C. Percival, and A. U. Tothill. 1970. *British Journal of Pharmacology* 40:161P–162P.
38. Castleman, M. 1991. *The healing herbs,* 294–296. Emmaus, PA: Rodale Press.
39. Bricklin, M. 1982. *Encyclopedia of natural home remedies,* 379–381. Emmaus, PA: Rodale Press.
40. Harvey, R. 1995/1996. *British Journal of Phytotherapy* 4:55–65.
41. Winterhoff, H., H.-G. Gumbinger, U. Vahlensieck, F. H. Kemper, H. Schmitz, and B. Behnke. 1994. *Arzneimittel-Forschung* 44:41–45.
42. Weiss, R. F. 1988. *Herbal medicine,* 275–278. Gothenburg, Sweden: AB Arcanum.
43. Yongchaiyudha, S., V. Rungpitarangsi, N. Bunyapraphatsara, and Choke-chaijaroenpron. 1996. *Phytomedicine* 3 (3): 241–243.
44. Khan, A., M. Safdar, M. Khan, K. N. Khattak, and R. A. Anderson. 2003. *Diabetes Care* 26 (12): 3215–3218.
45. Gupta, A., R. Gupta, and B. Lal. 2001. *Journal of the Association of Physicians in India* 49:1057–1061.
46. Mani, U. V., S. Desai, and U. Iyer. 2000. *Journal of Neutraceuticals, Functional & Medical Foods* 2 (3): 25–32.
47. Vuksan, V., J. L. Sievenpiper, V. Y. Y. Koo, T. Francis, U. Beljan-Zdravkovic, Z. Xu, and E. Vigden. 2000. *Archives of Internal Medicine* 160:1009–1013.
48. Vuksan, V., M. P. Stavrp. J. L. Sievenpiper, U. Beljan-Zdravkovic, L. A. Leiter, R. G. Josse, and Z. Xu. 2000. *Diabetes Care* 23:1221–1226.
49. Vuksan, V., M. P. Stavro, J. L. Sievenpiper, V. Y. Y. Koo, E. Wong, U. Beljan-Zdravkovic, T. Francis, A. L. Jenkins, L. A. Leiter, R. G. Josse, and Z. Xu. 2000. *Journal of the American College of Nutrition* 19:738–744.
50. Vuksan, V., J. L. Sievenpiper, J. Wong, Z. Xu, U. Beljan-Zdravkovic, J. T. Arnason, V. Assinewe, M. P. Stavro, A. L. Jenkins, L. A. Leiter, and T. Francis, T. 2001. *American Journal of Clinical Nutrition* 73:753–758.
51. Park, J. D., D. K. Rhee, and Y. H. Lee. 2003. *Phytochemistry Reviews* 4:159–175.
52. Sievenpiper, J. L., J. T. Arnason, L. A. Leiter, and V. Vuksan. 2002. *European Journal of Clinical Nutrition* 56:1–6.
53. Xie, J. T., Y.-P. Zhou, L. Dey, A. S. Attele, J. A. Wu, M. Gu, K. S. Polonsky, and C.-S. Yuan. 2002. *Phytomedicine* 9:254–258.
54. Attele, A. S., Y.-P. Zhou, J.-T. Xie, J. A. Wu, L. Zhang, L. Dey, W. Pugh, P. A. Rue, K. S. Polonsky, and C.-S. Yuan. 2002. *Diabetes* 51:1851–1858.
55. Sotaniemi, E. A., E. Haapakoski, and A. Rautio. 1995. *Diabetes Care* 18:1373–1375.
56. Vuksan, V., and J. L. Sievenpiper. 2005. *Nutrition Metabolism and Cardiovascular Diseases* 15:149–160.

chapter nine

Arthritic and musculoskeletal disorders

Arthritis

Arthritis refers to a whole spectrum of disorders, all of which are characterized by inflammation and tissue damage at the joints. Some of the various types are rheumatoid arthritis, juvenile arthritis, ankylosing spondylitis, psoriatic arthritis, Reiter's syndrome, osteoarthritis, and fibrositic disorders.

Of these, osteoarthritis is the most prevalent, affecting nearly 10 percent of the population older than the age of sixty. It ranks second only to cardiovascular disease in causing chronic disability because it involves the weight-bearing joints of the body, causing pain, limitation of motion, deformity, and progressive deterioration. Both sexes tend to be equally affected, and repetitive use of a particular joint through either work or leisure activities has been implicated in causing osteoarthritis.[1] Rheumatoid arthritis occurs in about 1–2 percent of the adult population and two to three times more often in women than in men. Epidemiologic data suggest that a genetic predisposition and exposure to unknown environmental factors may be necessary for expression of the disease. In addition, its cause is thought by many to be an infection, although years of searching have not revealed a causative organism.

Regardless of the type of arthritis, the immune response plays a significant role in producing both local inflammation and tissue damage. Macrophages engulf and process antigens that are then presented to T lymphocytes, where they stimulate the production of activated T cells. The activated T cells produce cytotoxins and cytokines, which stimulate the inflammatory processes and attract cells to areas of inflammation. Vasoactive substances (namely, histamine, kinins, and prostaglandins) are released at the site of inflammation, where they increase blood flow and permeability of blood vessels. These substances are responsible for the edema, warmth, erythema, and pain associated with inflamed joints.[2]

Initial drug therapy of most of the types of arthritis involves the systemic administration of salicylates or other nonsteroidal anti-inflammatory

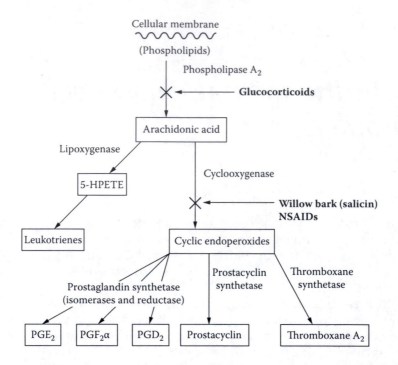

Figure 9.1 Pathways of biosynthesis for eicosanoids and sites of inhibition for anti-inflammatory drugs.

drugs (NSAIDs). Administered in sufficient doses, these compounds reduce inflammation, provide a degree of analgesia, maintain joint mobility, and help prevent deformity. They do not, however, alter the long-term progression of the disease. A major aspect of the mechanism of action of the NSAIDs is their ability to reduce prostaglandin biosynthesis through inhibition of cyclo-oxygenase. As illustrated in Figure 9.1, arachidonic acid, which is released from cell membrane phospholipids by the action of phospholipase A_2, is a central intermediate in the biosynthesis of eicosanoid compounds including the proinflammatory leukotrienes and prostaglandins. The NSAIDs inhibit the prostaglandin branch of the biosynthesis.

Another group of compounds important in the treatment of inflammatory disorders includes the glucocorticoids. They inhibit both branches of eicosanoid biosynthesis by preventing the release of arachidonic acid from the membrane phospholipids through inhibition of phospholipase A_2.[3] Although aspirin or other NSAIDs are the agents of choice for treating arthritis, it is important to mention that salicin, an active constituent of willow bark, has served as the prototype in the drug development of aspirin and the other NSAIDs (Chapter 1, Figure 1.1); consequently, willow bark tea has been used for the treatment of arthritis.

Herbal remedies for arthritis

Willow bark

About three hundred different species of the genus *Salix* are called willow. The one that is generally recognized as a source of medicinal bark in the United States is *Salix alba* L., but for reasons that will become clear, the bark of *Salix purpurea* L. and *Salix fragilis* L. are of superior quality.

The principal active constituent of willow bark was long thought to be a compound known as salicin, which chemically is salicyl alcohol glycoside. However, recent studies have shown that a whole series of phenolic glycosides designated salicortin (normally the main active principle in willow species), fragilin, tremulacin, etc. is present in the bark, some in much larger amounts than true salicin.[4] The glycosides, other than salicin, are relatively heat labile and are converted to the latter compound if the bark is dried at high temperature.[5] All of the phenolic glycosides have similar physiological effects, being prodrugs that are converted to the active principle, salicylic acid, in the intestinal tract and the liver.[6] Because of the time required for this conversion, the therapeutic properties of willow bark are expressed more slowly but continue to be effective for a longer time than if salicylate itself were administered.[7] Willow bark and its active constituents produce anti-inflammatory, analgesic, and antipyretic effects through their ability to inhibit cyclo-oxygenase and prostaglandin biosynthesis (see Figure 9.1).

For analytical purposes, the numerous phenolic glycosides are first converted to salicin and their content is then indicated as total salicin per unit weight of dried bark. Such studies show enormous variability, not only among different species of willow bark but also among different collections of the same species. Barks of high quality, such as those of *S. purpurea*, range from about 6–8.5 percent total salicin; one sample of *S. fragilis* even exceeded 10 percent. Probably a figure of 7 percent for such quality barks is average. However, none of the *S. alba* samples exceeded 1 percent, and other willow species had even less. In Germany, a standard of not less than 1 percent total salicin has been established for willow bark. This is indicative of the low level of total salicin in most commercial samples.[8]

Most of the willow bark available commercially is in the form of rather coarse pieces. Studies have shown that normal preparation of a tea from such material using hot water would extract only about 75 percent of the active principles. If a very finely powdered bark is used, the extraction will approach 100 percent of the activity.

The usual daily dose of aspirin for arthritic disorders is 3.6–5.4 g (average: 4.5 g) administered in divided doses. Equivalent amounts of other salicylates are also effective. The question then is concerned with how much total-salicin-containing willow bark must be administered to produce that amount of salicylate in the body. For purposes of this example, let us assume use of a relatively good-quality bark containing 7 percent total salicin. Let us also assume that the bark is finely powdered and is carefully extracted with sufficient hot water to obtain 100 percent of the active principles. Because the exact composition of the mixture of phenolic glycosides (salicin, salicortin, etc.) varies in the individual barks, it is not possible to calculate the exact amount of active salicylate produced from a given quantity of an unknown mixture. Theoretically, it will be considerably less than 50 percent, and that conversion will occur over several hours rather than immediately. But for convenience in calculating, let us assume 50 percent.

Based on these generous assumptions, it would be necessary to consume the tea prepared from about 130 g of bark to yield an average daily dose of salicylate sufficient to treat arthritic–rheumatic disorders. At the standard strength of 1 teaspoonful per cup of water, that is more than 6 L (1½ gal) of willow tea daily. Considering the high concentration of tannin (8–20 percent) in the bark and the sheer volume of liquid involved, consumption of this quantity is not feasible. If ordinary willow bark with its approximately 1 percent total salicin were to be substituted for the superior bark, these figures must be multiplied by seven, yielding 2 lb of bark and some 11 gal of tea—an utterly impossible daily dose. The need for such high dosage levels also precludes effective use of the crude herb in another form, such as capsules or tablets.

What about the use of willow bark for other conditions—headache, fever, sprains, strains, etc.—that respond to treatment with salicylates? Here, there will likely be confusion about proper dosage because the German Commission E recommends use of a quantity of bark equivalent to 60–120 mg of total salicin daily. Such a minimal dose would have little therapeutic value and seems to be based more on the quantity of bark used to prepare a normal cup of tea than upon any proven value of the small amount of total salicin contained therein. It does also point out that willow bark is used in German medicine as an adjuvant or auxiliary drug, rather than as an agent that in itself possesses great therapeutic value. In practice, it is supplemented by administration of synthetic salicylates.[7,9]

Side effects have not been reported with the use of willow bark tea, with the exception of gastrointestinal upset caused by the tannins present (not salicin). Precautions associated with salicylate therapy, such as hypersensitivity, are also applicable to willow bark; however, apparently because of the small amount of salicin present in the herbal tea, problems with side effects appear to be slight.[10]

The *National Formulary VI* listed the average dose of salicin as 1 g.[11] This is a reasonable analgesic–antipyretic dose, considering that the compound has less than half the activity of salicylates. It would require approximately 14 g (1/2 oz or 3 heaping teaspoonfuls) of high-quality willow bark (7 percent total salicin) prepared as three cups of tea to yield a single average dose of salicin. If ordinary white willow bark were employed, these quantities would have to be multiplied by seven.

In spite of the presence in various willow barks of total salicin—a mixture of therapeutically useful salicylate precursors—these herbs, by themselves, do not constitute effective treatment for arthritis or even for headaches, fevers, or muscle pains. The quantity of active principles present in even the highest-quality barks is insufficient to allow them to be consumed in sufficient amounts to constitute useful medicines. They may play a psychologically supportive role as auxiliaries to synthetic salicylate therapy in certain patients.

Feverfew

It has been claimed that people in the United Kingdom self-medicate with feverfew more for arthritis than for migraine[12]; however, the single clinical trial in rheumatoid arthritis failed to show any beneficial effect in forty women treated with 70–86 mg of dried feverfew leaf or placebo for six weeks.[13] Notwithstanding, it has been suggested that feverfew may well be of benefit in milder cases of arthritis than that which afflicted the women in this trial, who were extremely refractory cases unresponsive to all conventional drugs.[14] The authors of this failed trial have suggested, particularly, that feverfew may be of benefit in osteoarthritis and soft tissue lesions, for which self-treatment with feverfew is probably most common.[15]

Muscle pain

External analgesics, many of which are of plant origin, are widely used to allay the discomfort associated with the overuse of skeletal muscles. When stimulated by strenuous exercise or other irritations, pain receptors located in the skeletal muscles transmit impulses to the brain that are interpreted as pain. Such deep-seated pain is difficult to characterize; it is often described as "dull" or "aching."

It is postulated that a neural mechanism in the spinal cord acts like a gate to control the transmission of pain to the brain. Pain signals are carried from pain receptors in the muscles to the spinal cord via two types of nerve fibers designated A delta and C. Type C fibers are small and unmyelinated, conduct impulses slowly, and are associated with dull, aching pain. Type A delta fibers are large, contain myelin, and are linked with immediate pain that is sharp and stabbing. The two types of nerve fiber impulses can oppose each other, and mild stimulation of type A delta fibers can attenuate pain felt from the activation of type C fibers. The theory is that stimulation of the type A delta fibers closes a gate in the spinal cord so that the pain impulses carried by the type C fibers are not transmitted to the brain.[16]

External analgesic agents are used not only to alleviate pain resulting from trauma to muscles due to sports injuries and workplace injuries, but also to treat the pain that results from arthritis. As a consequence, these agents have a wide use with estimates that 40 percent of adults over the age of fifty are regular users of such products; arthritis patients account for more than 75 percent of total usage.

External analgesics are also known as counterirritants because, when they are applied to the skin over or near the site of underlying pain, they produce a mild, local irritation that stimulates type A delta nerve fibers; this acts to counter the dull, throbbing pain transmitted by the type C nerve fibers. A classification of counterirritants is based on the different characteristics of their physiologic action. Those that produce dilation of the cutaneous vasculature and subsequent reddening of the skin are called rubefacients. The resulting localized reaction of warmth and redness causes the patient to disregard the original deep-seated pain. In general, these agents are more potent than other counterirritants, and exposure to the skin must be carefully controlled because their strong irritation may cause erythema and blistering.

A second category of counterirritants is those that produce a cooling sensation on the skin. They depress sensory cutaneous receptors and act as local anesthetics or antipruritics at lower concentrations (0.1–1 percent). In higher concentrations, the initial feeling of coolness is followed by a sensation of warmth. The last category includes agents that produce irritation and warmth to the skin but without producing rubefaction.[16]

Rubefacients (agents that induce redness and irritation)

Volatile mustard oil (allyl isothiocyanate)

This is a volatile oil obtained from the dried, ripe seed of varieties of black mustard, *Brassica nigra* (L.) W. D. J. Koch, or of *B. juncea* Czern. & Coss, family Brassicaceae.[17] The seed is ground and macerated in water to allow the enzyme myrosin to convert the glycoside sinigrin to allyl isothiocyanate. This volatile compound is then purified by distillation and incorporated into various counterirritant preparations intended for external application.

Alternatively, a poultice may be prepared from equal parts of the powdered seeds and flour by moistening with sufficient water to form a paste. The mixture is spread on a cloth and applied to the affected area for a short period of time. If left too long, the continued release of the volatile mustard oil may blister the skin.

Although it is extremely acrid and irritating, volatile mustard oil is considered by the FDA to be a safe and effective counterirritant if applied in concentrations ranging from 0.5 to 5.0 percent. It may be used as frequently as three or four times daily.

Methyl salicylate

This compound is obtained by distilling the leaves of wintergreen, *Gaultheria procumbens* L. (family Ericaceae), or the bark of the sweet birch, *Betula lenta* L. (family Betulaceae). These products are referred to as wintergreen oil and sweet birch oil, respectively. They are essentially indistinguishable from each other and from methyl salicylate that is prepared synthetically by the esterification of synthetic salicylic acid.[18]

Methyl salicylate is applied topically as a counterirritant in the form of liniments, gels, lotions, or ointments containing concentrations of 10–60 percent. The number of applications should not exceed three or four per day. Strenuous physical activity and heat increase the percutaneous absorption of methyl salicylate, which may result in salicylate toxicity. Consequently, users should be warned not to apply methyl salicylate after vigorous exercise in hot and humid weather or to follow up the application by use of a heating pad.[16]

It has been reported that a considerable amount of methyl salicylate may be absorbed through the skin after topical application. The absorption is increased with multiple applications. After applications to the thigh of 5 g of a 12.5 percent methyl salicylate ointment twice daily for four days, the recovery of total salicylates (salicylic acid and its principal metabolites) in the urine from twelve subjects averaged 22 percent on day four. This level of salicylate would warrant caution in patients for whom systemic salicylate may be hazardous or problematic.[19]

Turpentine oil

Sometimes called spirits of turpentine, this is a volatile oil distilled from the oleoresin obtained from the long-leaf pine, *Pinus palustris* Mill., family Pinaceae. It has a long history of use in counterirritant preparations and is employed in concentrations of 6–50 percent. Application should not exceed three or four times daily.

Refrigerants (agents that induce a cooling sensation)

Two additional products are slightly less effective than the counterirritants previously discussed. Nevertheless, they are widely used. Both differ from the previous ones in that they produce a strong cooling sensation when applied to the skin.

Menthol

This is an alcohol obtained from various mint oils or produced synthetically. It is employed three or four times a day in topical preparations in concentrations of 0.1–1.0 percent as an antipruritic. When used in higher concentrations of 1.25–16 percent, it acts as a counterirritant. The effect of menthol on thermal sensations, pain, experimental itch, and irritation was studied in eighteen subjects. The results suggest that menthol fulfills the definition of a counterirritant but has little effect on histamine-induced itch and does not affect pain sensation as a local anesthetic.[20]

Camphor

Camphor is a ketone obtained from *Cinnamomum camphora* (L.) J. S. Presl of the family Lauraceae or produced synthetically. It is used three or four times daily in topical preparations containing concentrations of 0.1–3.0 percent, depressing cutaneous receptors and providing topical analgesic, local anesthetic, and antipruritic effects. The recommended concentration for topical use as a counterirritant for adults and children more than two years of age is 3–11 percent.

Other counterirritants

The *Capsicum*-derived capsaicin cream discussed in Chapter 7 as a pain controller also provides counterirritant activity without rubefaction. For details of its composition and use, see the discussion there.

Gout

Gout is a recurrent form of acute arthritis resulting from a deposition in and around the joints and tendons of crystals of monosodium urate. It is extremely painful and, if not treated, can lead to chronic disability.[21]

Herbal remedy for gout

Colchicum

One of the most ancient treatments for gout is colchicum. For more than five hundred years, the corm of the autumn crocus, *Colchicum autumnale* L. of the family Liliaceae, has been used as a specific treatment to alleviate the inflammation associated with gout. Then, about 175 years ago, colchicum seeds began to be utilized for the same purpose. In 1820, the active principle, colchicine, was first isolated, and in the intervening years, it has replaced the herb (corm and seed) as the dosage form of choice.

In therapeutic doses, colchicine displays a number of unpleasant side effects; in large doses, it is quite toxic.[22] As a result, neither it nor the plant material containing it is available for over-the-counter acquisition. It will not be discussed further here.

References

1. Boh, L. E. 1997. Osteoarthritis. In *Pharmacotherapy: A pathophysiologic approach*, 3rd ed., ed. J. T. DiPiro, R. L. Talbert, G. C. Yee, G. R. Matzke, B. G. Wells, and L. M. Posey, 1735–1753. Stamford, CT: Appleton & Lange.
2. Schuna, A. A., M. J. Schmidt, and D. W. Pigarelli. 1997. Rheumatoid arthritis and the seronegative spondyloarthropathies. In *Pharmacotherapy: A pathophysiologic approach*, 3rd ed., ed. J. T. DiPiro, R. L. Talbert, G. C. Yee, G. R. Matzke, B. G. Wells, and L. M. Posey, 1717–1733. Stamford, CT: Appleton & Lange.
3. Brestal, E. P., and K. Van Dyke. 1997. Anti-inflammatory and antirheumatic drugs. In *Modern pharmacology with clinical applications*, 5th ed., ed. G. R. Craig and R. E. Stitzel, 455–469. Boston: Little, Brown and Company.
4. Meier, B., O. Sticher, and A. Bettschart. 1985. *Deutsche Apotheker Zeitung* 125:341-347.
5. Julkunen-Tiitto, R., and K. Gebhardt. 1992. *Planta Medica* 58:385–386.
6. Meier, B., and M. Liebi. 1990. *Zeitschrift für Phytotherapie* 11:50–58.
7. Schneider, E. 1987. *Zeitschrift für Phytotherapie* 8:35–37.
8. Bisset, N. G., ed. 1994. *Herbal drugs and phytopharmaceuticals*, English ed. (*Teedrogen*, M. Wichtl, ed.), 437–439. Boca Raton, FL: CRC Press.
9. Hänsel, R., and H. Haas. 1984. *Therapie mit Phytopharmaka*, 233–235. Berlin: Springer–Verlag.
10. Newall, C. A., L. A. Anderson, and J. D. Phillipson. 1996. *Herbal medicines: A guide for health-care professionals*, 268–269. London: The Pharmaceutical Press.
11. *The national formulary*, 6th ed. 1935. Washington, D.C.: American Pharmaceutical Association, 322–323.
12. Johnson, E. S. 1983. *MIMS Magazine* May 15: 32–35.
13. Pattrick, M., S. Heptinstall, and M. Doherty. 1989. *Annals of the Rheumatic Diseases* 48:547–549.
14. Heptinstall, S. Personal communication to D. V. C. Awang, 1990.
15. Heptinstall, S., and D. V. C. Awang. 1998. *Phytomedicines of Europe: Chemistry and biological activity*, 158–175. ACS Symposium Series 691. Washington, D.C.: American Chemical Society.
16. Jacknowitz, A. I. 1996. External analgesic products. In *Handbook of nonprescription drugs*, 11th ed., ed. T. R. Covington, 75–90. Washington, D.C.: American Pharmaceutical Association.
17. Robbers, J. E., M. K. Speedie, and V. E. Tyler. 1996. *Pharmacognosy and pharmacobiotechnology*, 59–60. Baltimore, MD: Williams & Wilkins.
18. Osol, A., and G. E. Farrar, Jr. 1947. *The dispensatory of the United States of America*, 24th ed., 707–708. Philadelphia, PA: J. B. Lippincott.
19. Morra, P., W. R. Bartle, S. E. Walker, S. N. Lee, S. K. Bowles, and R. A. Reeves. 1996. *Annals of Pharmacotherapy* 30:935–940.
20. Yosipovitch, G., C. Szolar, X. Y. Hui, and H. Maibach. 1996. *Archives of Dermatological Research* 288:245–248.
21. Hawkins, D. W., and D. W. Rahm. 1997. Gout and hyperuricemia. In *Pharmacotherapy: A pathophysiologic approach*, 3rd ed., ed. J. T. DiPiro, R. L. Talbert, G. C. Yee, G. R. Matzke, B. G. Wells, and L. M. Posey, 1755–1761. Stamford, CT: Appleton & Lange.
22. Van Dyke, K. 1997. Drugs used in gout. In *Modern pharmacology with clinical applications*, 5th ed., ed. C. R. Craig and R. E. Stitzel, 471–478. Boston: Little, Brown and Company.

chapter ten

Problems of the skin, mucous membranes, and gingiva

Dermatitis

Dermatitis is not so much a disease as it is a symptom—an erythematous condition (redness) of the skin characterized by inflammation. It may result from various external or internal causative factors. Dermatitis may be acute or chronic. It is common to observe vesicles on a reddish base along with oozing (weeping) of fluid in acute dermatitis. In chronic dermatitis, the weeping subsides and the skin becomes dry and scaly with the appearance of fissures in the dermis. Secondary infections may occur as a consequence of pruritis-induced scratching.

The most common forms of dermatitis include contact dermatitis and atopic dermatitis, also called eczema, occurring primarily in infants, children, and young adults. Irritant contact dermatitis results from exposure of the skin to chemicals and does not require allergic sensitization. Examples of causative agents are detergents and household cleansers. Allergic contact dermatitis is a delayed hypersensitivity reaction to contact allergens. An initial sensitizing exposure to the allergen is required; upon subsequent re-exposure to it, the dermatitis will occur. The most common example of this type of dermatitis is the rash that occurs after exposure of the skin to the leaves, stems, and roots of poison ivy.[1]

Herbal treatment of dermatitis is a conservative therapy involving, first of all, removal of the irritating or allergenic agent from the environment. Soaps and detergents are primary causative agents if the hands are involved. This is followed by application of a solution with astringent properties to reduce the weeping. Such solutions generally function by coagulating the surface proteins of the cells, thus reducing their permeability and limiting the secretion of the inflamed tissue. The precipitated proteins also tend to form a protective layer, thereby limiting bacterial development and facilitating the growth underneath of new tissue. Tannin-containing herbs are especially effective astringents.

Tannin-containing herbs

Witch hazel leaves

One of the most widely used herbs in this category is witch hazel leaves. Also known as hamamelis leaves, this herb consists of the dried leaves of *Hamamelis virginiana* L., a native American shrub or small tree of the family Hamamelidaceae. The leaves contain 8–10 percent of a mixture of tannins consisting chiefly of gallotannins with some condensed catechins and proanthocyanins.[2]

An aqueous or hydroalcoholic extract of witch hazel leaves contains large amounts of tannin and is an excellent astringent. Unfortunately, such preparations are not ordinarily available in the United States. The only common commercial product here is technically referred to as hamamelis water, witch hazel extract, or distilled witch hazel extract; however, in the vernacular it is known simply as witch hazel.

Distilled witch hazel extract is prepared by macerating in water and then steam distilling the recently cut, dormant twigs of the plant. Alcohol is then added to the distillate to obtain a final concentration of 14–15 percent. The FDA has declared this product, prepared according to the procedure described in the *National Formulary XI* in 1960, under the colloquial designation "witch hazel," to be a safe and effective astringent. Because the tannins are not carried over into the extract during the distillation process, the only plant constituent present is a trace of volatile oil too limited in amount to exert any therapeutic influence. Whatever astringent activity is present—and that must be very limited indeed—can only be attributed to the 14–15 percent alcohol contained in the product.

Authentic (nondistilled) hydroalcoholic extracts of witch hazel leaves are, on the other hand, much more effective astringents and styptics. They are used as such or following incorporation into an ointment base to relieve local inflammation of the skin and mucous membranes and to treat hemorrhoids and varicose veins.

Witch hazel has a low toxicity profile, but internal use is not recommended.[3] A decoction can be prepared for external use as a poultice by using 5–10 heaping teaspoonfuls of finely chopped leaves to a cupful (240 mL) of water. Bring to a boil and simmer for five to ten minutes, and then strain.

Oak bark

When this product was an official drug in the *National Formulary*, it was said to consist of the dried inner bark of the trunk and branches of the white oak, *Quercus alba* L., family Fagaceae. However, the barks of a number of different oaks have been used medicinally, including *Q. robur* L., the British oak, and *Q. petraea* (Matt.) Liebl., the winter oak.

Oak bark has a highly variable total tannin content of 8–20 percent, which includes catechins, oligomeric proanthocyanidins, and ellagitannins. For external use, a decoction is prepared from 2 teaspoon-fuls of the coarsely powdered bark and 500 mL of water. After straining, this aqueous extract is applied directly to the affected skin, including the genital and anal regions of the body; it is also used as a mouthwash for inflammation of the mucous membranes of the oral cavity.[4]

English walnut leaves

The dried leaves of the English walnut, *Juglans regia* L., family Juglandaceae, contain about 10 percent of an ellagic acid–derived tannin and are there-fore utilized as a local astringent in a manner similar to that for the pre-ceding herbs. A decoction prepared by boiling 5 teaspoonfuls of the leaves in 200 mL of water and straining is applied to the affected skin three or four times a day.[5]

Other herbal products

As noted previously, a number of ointments, creams, and lotions contain-ing chamomile volatile oil and intended for the treatment of various skin conditions are currently marketed in Europe. Because these pharmaceuti-cals have not been approved for sale in the United States and consequently are not generally available there, they are not discussed in this chapter. See Chapter 3 for a detailed discussion of this herb. Evening primrose oil (discussed in Chapter 8) has been thought to be of value in treating atopic dermatitis (eczema). However, conclusive proof of its value in such condi-tions is lacking.

Contact dermatitis

The most severe type of allergic contact dermatitis is that produced by touching poison ivy, *Toxicodendron radicans* (L.) O. Kuntze or other related species of *Toxicodendron* known as poison oak or poison sumac. A toxic mixture of catechols designated urushiol contained in this species serves as the allergen. Following initial sensitization, it produces an acute contact dermatitis characterized by weeping, vesicular lesions with edema, crust-ing, and, occasionally, secondary infection. Itching may be severe. The time between contact with the allergen and the first sign of reaction is usu-ally two to three days. Most cases of the dermatitis are self-limiting and disappear in fourteen to twenty days. Conventional treatment, depending on the severity of the condition, involves the systemic or local administra-tion of corticosteroids and drying lotions applied to the lesions.[6]

More than one hundred different plants or plant products have been used in the past to treat poison ivy. These include the counterirritants menthol and camphor that, when applied topically in low concentration, relieve irritation by the depression of cutaneous receptors. Astringent herbal products such as witch hazel stop weeping, reduce inflammation, and promote healing of the dermatitis. Another herb that has demonstrated effectiveness is jewelweed.

Herbal remedy for contact dermatitis

Jewelweed

In a study of Indiana folk medicine, the use of jewelweed, *Impatiens biflora* Walt. or *I. pallida* Nutt. (family Balsaminaceae), taken internally or applied externally, was repeatedly recommended as a cure for poison ivy dermatitis. Internally, a decoction prepared from any part of the plant is used; externally, the sap of the stem is applied to the affected area.[7]

The results of a clinical study in which a 1:4 jewelweed preparation was compared for its effectiveness with other standard poison ivy dermatitis treatments were published in 1958.[8] Of the 115 patients treated with jewelweed, 108 responded "most dramatically to the topical application of this medication and were entirely relieved of their symptoms within two or three days after the institution of treatment." It was concluded that jewelweed is an excellent substitute for ACTH and the corticosteroids in the treatment of poison ivy dermatitis. The active principle in the plant that is responsible for this activity remains unidentified.

Burns, wounds, and infections

Because a single herbal remedy is often used to treat two or more of these conditions, they are all considered in a single section. Burns and wounds require treatment to alleviate pain, to prevent infection, and to facilitate regeneration of tissue. Both occur quite commonly, especially in the home. For example, 80 percent of burns occur there, but only 5 percent are severe enough to require hospitalization. Of the minor burns occurring outside the home, sunburn is the most common. The goals of treating minor burns (those not affecting full-skin thickness) are to relieve the pain, protect the burn from air, prevent dryness, and provide a favorable environment for healing without infection.[9]

Abrasions or minor cuts or wounds are normally cleansed using mild soap and water to remove debris that might lead to infection, and an antiseptic preparation is applied. Ideally, such an antiseptic should be minimally affected by the presence of organic materials. Such preparations often stain the tissue. Although this is generally undesirable, it may be useful in delineating a "clean" area. More serious or deeper tissue infection (e.g., puncture wounds and severe burns) requires consultation with a physician to determine if systemic or topical antibiotic therapy is required.[10]

A relatively large number of superficial fungal infections of the keratinized tissue of the skin, hair, and nails occur in humans. They may be chronic and resistant to treatment; however, they rarely affect the general health of the patient. Infections include ringworm of the scalp or body, jock itch, athlete's foot, candidiasis, and the like.[11] Normally, such conditions are treated by the topical application of antifungal agents, although in stubborn cases it may be necessary to resort to the systemic administration of appropriate antifungal drugs.

In recent years, the incidence of so-called vaginal yeast infections has increased somewhat, but public awareness of the condition has increased markedly. The common causative agent is not a true yeast but, rather, *Candida albicans* (or a related *Candida* species), an imperfect fungus with a yeast-like appearance. The increased incidence is attributed to more widespread use of broad spectrum antibiotics, which permit *Candida* infections to develop by destroying the normal vaginal flora, and to the popularity of occlusive female clothing, particularly pantyhose. Public awareness has been promoted by extensive advertising of two synthetic vaginal fungicides, clotrimazole and miconazole, which were switched to OTC status in 1991.

Herbal remedies for burns, wounds, and infections

Aloe gel

One of the most widely used herbal preparations for the treatment of various skin conditions is aloe gel, also termed aloes in pharmacy. Often referred to as aloe vera gel, this is the mucilaginous gel obtained from the cells making up the inner portion (parenchyma) of the leaf of *Aloë barbadensis* Mill. (family Liliaceae), sometimes referred to as *A. vera* (L.) Webb & Berth. or *A. vulgaris* Lam. The gel should not be confused with the bitter yellow latex or juice that derives from the pericyclic tubules occurring just beneath the epidermis or rind of the leaves. That material, in dried form, is a potent laxative and was discussed in Chapter 3.[12] It should be emphasized that "juice" in this context should be distinguished from the "juice" sold as a beverage, which is a thin liquid expressed from aloe gel.

Aloe gel has been widely used externally for its wound-healing properties and internally as a general tonic or cure-all. It is incorporated into a wide variety of ointments, creams, lotions, shampoos, and the like for external use. Although there appears to be general agreement regarding the effectiveness of fresh aloe gel in the treatment of minor skin ailments, considerable controversy surrounds the effectiveness of these aloe preparations.

In the first place, the gel used in the manufacture of these preparations is often obtained and treated in different ways. The original process for obtaining the gel was to remove the outer layers of the leaf including the pericyclic cells by filleting with a knife, leaving the inner core of gel. Recently, in order to eliminate some of the intensive hand labor the preceding process requires and to increase the recovery of gel solids, some manufacturers have started processing the whole leaf. Analysis of the components in the final products produced through the two processes showed that the concentration of certain components was very different. The polysaccharide concentration was much higher in the inner-gel product than in the whole-leaf product; therefore, the former may be preferred for medicinal applications,[13] even though it has not been firmly established which constituents of the gel are required for therapeutic activity. In addition, some of the gel on the market is "reconstituted" from a powder or concentrated liquid. "Aloe vera extract" usually denotes a diluted product.[14] These differences in processing and the resulting composition of the product have resulted in differences in observed activity of the commercial products.

The constituents of aloe gel have been studied in some detail, with the bulk of the gel being a mucilage of a polysaccharide nature, constitutiong 0.2–0.3% of the fresh gel.[15] The polysaccharide components can be grouped into three main categories: the partially acetylated β-1,4-glucomannans; the neutral polysaccharides, including a galactan, an arabinan, and a non-acetylated glucomannan; and acidic galacturonans (similar to pectin).[16] The aloe glucomannans are closely related structurally to cellulose, and both polysaccharides are cleaved or digested by the enzyme cellulase. Therefore, if cellulase is used in the gel production process to reduce slime, the composition of the final product will show an increase in small saccharides, which may make for an inferior product.[13]

One study of the effect of aloe gel on the growth of human cells in artificial culture found that the fresh material promoted significantly the attachment and growth of the cells. It also enhanced the healing of wounded monolayers of the cells. However, a "stabilized" commercial product actually proved toxic to such cells, disrupting their attachment and growth.[17] Recently, these findings were investigated further using gel permeation column chromatography to obtain fractions of various components of aloe gel. These were tested by in vitro assays for proliferation of human normal dermal and baby hamster kidney cells. The glycoprotein fraction promoted cell growth, while the neutral polysaccharide fraction did not show any growth stimulation. Moreover, a colored glycoprotein fraction strongly inhibited the in vitro assays. The investigators speculated that the conflicting activities between the proliferative glycoprotein and the inhibitory, colored glycoprotein may explain the variability in the results of pharmacological and therapeutic experiments with aloe gel.[18]

Other studies have shown that aloe gel and some preparations containing it were useful in the treatment of various types of skin ulceration in humans and burn and frostbite injuries in animals.[19] In another study, a cream base containing aloe was shown to be effective in preserving circulation in the skin after frostbite injury.[20] Stabilized aloe vera was shown to produce a dramatic acceleration of wound healing in patients who had undergone full-face dermabrasion.[21] In a study of twenty-seven patients with partial thickness burn wounds, those treated with aloe gel healed faster (average of twelve days) than those treated with Vaseline gauze (average of eighteen days).[22] A double-blind, placebo-controlled study involving sixty patients found that 0.5 percent aloe gel extract in a hydrophilic cream was more effective than placebo in treating psoriasis vulgaris.[23] Thus, evidence is beginning to accumulate to support not only the effectiveness of fresh aloe gel but also some preparations containing the processed product.

Whereas the active constituents of aloe gel have not yet been identified, it is presently believed that some of its beneficial effects apparently result from the ability of a contained carboxypeptidase to inhibit the pain-producing agent bradykinin. The product is also believed to hinder the formation of thromboxane, the activity of which is detrimental to burn wound healing. Antiprostaglandin activity has also been reported. Aloe gel has antibacterial and antifungal properties, but little is known about the identity and stability of the ingredients responsible for these or most of its other effects. Many of the active constituents undoubtedly deteriorate upon storage, so use of the fresh gel is the only way to ensure maximal activity. Although the FDA concluded that there was insufficient evidence to support the effectiveness of aloe gel in treating any condition,[24] the product continues to be widely used as a remedy for minor skin ailments—cuts, bruises, abrasions, burns, etc.

Freshly prepared aloe gel is worthy of consideration as a home remedy to treat minor burns. *A. barbadensis* can easily be cultivated as a house plant. It likes plenty of light but not direct sun, and because it flourishes in arid climates, care must be taken not to overwater the plant. For treatment, a leaf can be removed from a plant and split open longitudinally, and the thin, clear gel may be scooped off the split leaf blade with a spoon.

Arnica

The dried flower heads of *Arnica montana* L. (family Asteraceae) and several other related species of *Arnica* have a long-standing reputation as a useful treatment for acne, bruises, sprains, muscle aches, and as a general topical counterirritant. The plant contains a number of sesquiterpene lactones, including helenalin, dihydohelenalin, arnifolin, and the arnicolides; flavonoid glycosides; and about 0.3 percent of a volatile oil.[25] The essential active principles are the helenalin and dihydrohelenalin esters, which have been shown to have strong antimicrobial, antiedema, and anti-inflammatory properties. Unfortunately, helenalin and its derivatives are also allergenic and may induce topical dermatitis.[26]

In the United States, arnica is commonly employed in the form of a tincture (hydroalcoholic extract) that is applied externally to the area to be treated. In Europe, arnica-containing creams are especially popular. The herb is approved by German Commission E for its anti-inflammatory, analgesic, and antiseptic properties for topical application.[27] It should not be taken internally.

Calendula

The flower heads of *Calendula officinalis* L. (family Asteraceae), commonly known as the garden marigold, have long been valued in Europe for the treatment of various skin ailments and to facilitate wound healing and reduce inflammation. The herb contains flavonoids, terpenoids, and a volatile oil, but the chemical principles responsible for the therapeutic effects remain unidentified.[28]

Calendula is utilized in the form of a tea prepared by pouring a cup of boiling water over 1–2 teaspoonfuls of herb and allowing it to steep for ten minutes. This preparation is used as a gargle or mouthwash for sores in the mouth or poured over an absorbent cloth and applied as a poultice to skin ailments. Occasionally, the tea is even consumed internally for its antispasmodic and other putative effects.[29] Extract of calendula is also incorporated into soaps as well as various ointments, creams, and sprays intended for local application.

German Commission E has found calendula to be effective for the reduction of inflammation and promotion of granulation of wounds.[30] In short, it promotes wound healing on local application.

Comfrey

This herb, also called common comfrey, should consist of the leaves or the root of *Symphytum officinale* L. of the family Boraginaceae; however, in commercial practice today it is seldom differentiated from prickly comfrey, *S. asperum* Lepech., or from Russian comfrey, *S.* x *uplandicum* Nym. Comfrey was initially applied externally to reduce the swelling around broken bones. Then it began to be used for treating a variety of wounds and, more recently, internally as a kind of tonic.[31]

Whatever therapeutic value comfrey may possess is attributed to its content of allantoin, a cell proliferant, and to rosmarinic acid, an anti-inflammatory agent and inhibitor of microvascular pulmonary injury.[32,33] Unfortunately, these beneficial principles are accompanied in the herb by a large number of toxic pyrrolizidine alkaloids that have been shown to cause cancer in small animals. Four cases of human poisoning by comfrey have now been reported, although it is not certain that all of them resulted from the ingestion of *S. officinale* (common comfrey). The identity of the species is important because it is now recognized that, although *S. officinale* contains toxic pyrrolizidine alkaloids (PAs), it usually does not contain large quantities of the highly toxic echimidine found in *S. asperum* (prickly comfrey) or *S. x uplandicum* (Russian comfrey).[34]

Originally, it was believed that *S. officinale* did not contain echimidine, but a 1989 chemotaxonomic study revealed that about one-fourth of the common comfrey samples examined did contain that alkaloid. This was generally in very small amounts, except for one $2n = 24$ cytotype, which contained levels of echimidine comparable to concentrations found in *S. asperum*.[35] It is still reasonable to conclude that a high level of echimidine in a comfrey sample indicates a species other than *S. officinale*.

Awang has shown that about half of the comfrey products available in Canada contained echimidine, in spite of the fact that they were labeled as consisting of common comfrey. This inattention to proper taxonomic identification by suppliers creates a particularly dangerous situation. In addition, comfrey roots contain about ten times the concentration of PAs found in leaves, rendering the roots unsuitable for any therapeutic application. Health and Welfare Canada has long refused to register any comfrey root products for medicinal application.

Reflecting its former enormous popularity, comfrey continues to be used both externally and internally by many people who are misinformed by advice provided in popular herbals. Believing that the use of herbs should be helpful, not harmful, to the consumer's health, we must conclude:

1. The internal use of any species of comfrey should be avoided.
2. Comfrey root should never be used medicinally.
3. Only the mature leaves of *S. officinale* should be applied externally and then only to intact skin for limited periods of time.
4. Comfrey should never be used by pregnant or lactating women or by young children.
5. Because so many other nontoxic yet effective treatments for minor skin ailments do not present the hazards associated with this herb, comfrey has little, if any, place in our modern materia medica.

In 1992, the German Federal Health Agency, in an attempt to balance the popularity of certain herbs containing toxic PAs with a recognized need for their restricted usage, established standards for such preparations. These limitations restricted the amount of total PAs with 1,2-unsaturated necine moieties that might be obtained from certain registered preparations to not more than 100 mg per day when used externally and 1 mg per day when taken internally. An exception was made for comfrey leaf tea, which was permitted to provide a maximum internal daily dose of 10 mg. In the case of nonregistered preparations, the limitations were reduced to 10 mg externally and 0.1 mg internally daily.[36]

It is necessary to point out that these limitations, which apply not only to comfrey but also to other herbs such as borage, coltsfoot, life root, and heliotrope, are meaningless in the United States because none of the preparations sold here has been assayed to determine its PA content. Further, no adjustment has been made in any of the available dosage forms to ensure that safe levels of toxic alkaloids are not exceeded. For this reason, safety requires adherence to the five rules stated earlier when dealing with comfrey. Similar precautions should be taken with all herbs containing toxic PAs.

Hydroalcoholic extracts of both *S.* x *uplandicum* herb and *S. officinale* root have been clinically proven effective in treatment of ankle distortions.[37,38] In the latter case, the proprietary comfrey ointment (Kytta-Salbe™, Merck GmbH, Darmstadt, Germany) was compared to the cyclo-oxygenase inhibitor diclofenac as gel (Vooltaren™, Novartis Pharmaceuticals, East Hanover, NJ) in eighty-two patients of mean age of twenty-nine years. Both treatments reduced the key symptoms related to ankle distortions, but on the basis of the AUC data and 95 percent confidence intervals, a statistically significant superiority of the comfrey extract was evident.

Tea tree oil

When distilled with steam, the leaves of an Australian tree, *Melaleuca alternifolia* (Maiden & Betche) Cheel of the family Myrtaceae, yield about 2 percent of a pale yellow oil known as tea tree oil. Tea tree oil is a complex mixture of more than one hundred components, predominantly monoterpenes and sesquiterpenes. The concentration of each component can vary widely depending on the oil sample; however, for commercial purposes, official Australian standards stipulate that the oil should contain up to a maximum of 15 percent 1,8-cineole and no less than a minimum of 30 percent 1-terpinen-4-ol. Other major components include *p*-cymene, linalool, α- and γ-terpinene, α-terpineol, and terpinolene. These components, plus 1,8-cineole and 1-terpinen-4-ol, constitute 80–90 percent of the oil.[39]

The oil was said to have been used by the aborigines as a local antiseptic, and early settlers in Australia utilized it for the treatment of cuts, abrasions, burns, insect bites, athlete's foot, and the like.[40] Because of the presence of large amounts of terpinen-4-ol, the tea tree oil does possess a pronounced germicidal activity. In fact, during World War II, it was incorporated in the machine "cutting" oils used in munitions factories in Australia to reduce the number of infections resulting from the metal filings and turnings that accidentally penetrated the skin of the workers' hands.[41]

A very few modern clinical studies have shown the possible value of tea tree oil in treating various vaginal and skin infections and acne vulgaris.[42–44] In earlier studies the antimicrobial activity of the oil was attributed entirely to 1-terpinen-4-ol. A recent study using a broth microdilution method for testing has shown that in addition to 1-terpinen-4-ol, a-terpineol and linalool have antimicrobial activity.[39]

Additionally, a recent study reported that sixty-four different isolates of methicillin-resistant *Staphylococcus aureus* were susceptible to tea tree oil. Using the broth microdilution method, the minimum inhibitory concentration and minimum bactericidal concentration values for sixty of the isolates were all 0.25 percent and 0.5 percent, respectively.[45] A multicenter, randomized, double-blind clinical trial compared the efficacy of tea tree oil to a topical 1 percent clotrimazole solution to treat toenail onychomycosis (fungal infection). Applied topically twice daily for six months to the affected nails of two groups of patients (117 total), the two treatments were found to be comparable in efficacy of cure.[46] Although irritation of the skin occasionally results in sensitive individuals, the use of the oil has not been associated with any particular toxicity.

Yogurt

This is a semisolid, curdled, or coagulated milk product resulting from the action of certain bacteria on the sugars in milk. Classically, two organisms are used in combination, *Lactobacillus bulgaricus* and *Streptococcus thermophilus*, which are allowed to act on whole fresh cow's milk at a temperature of 40–50°C for a period of five to ten hours.[47] In modern "low-fat" or "no-fat" yogurts, skim milk powder is added to milk with a reduced butterfat content to improve the consistency of the product, and the mixture is fermented with these bacteria or *Lactobacillus acidophilus*. Although yogurt is certainly not an herb, it may be thought of in this case as simply a substrate for the bacteria that are traditionally classified as microscopic plants. Consequently, it is the therapeutic utility of the living plants that is really being discussed here.

Results of a clinical study published in 1992 showed that the daily consumption for a six-month period of 8 oz (240 g) of yogurt containing live *L. acidophilus* greatly reduced the incidence of recurrent candidal vaginitis.[48] Hilton and colleagues postulated that the bacteria colonized the gastrointestinal tract and then the vaginal canal, where they restored or supplemented the normal lactobacilli flora and inhibited the growth of candida ("yeast").

As a result of these findings, the investigators recommended daily consumption of a cup of yogurt containing live *L. acidophilus* bacteria to decrease both candidal colonization and infection of the vagina. The problem is that not all yogurts contain live cultures of this species. Further, the claims of dairy products manufacturers regarding the method of preparation of their various yogurts are not always accurate. Apparently, those most likely to contain *L. acidophilus* are the low-fat rather than the no-fat varieties. Interested consumers can always make their own by following the directions provided for yogurt preparation in any reliable cookbook but using as a starter the contents of commercially available *L. acidophilus* capsules instead of a quantity of commercial yogurt. Alternatively, it seems reasonable to assume the consumption of the capsules themselves would produce the same results as the yogurt. Further clinical trials are definitely warranted.

Lesions and infections of the oral cavity and throat

Various pain-inducing abnormalities of the oral cavity are quite common. Some 20–50 percent of Americans suffer from canker sores (recurrent aphthous ulcers)—painful lesions apparently resulting from a dysfunction of the immune system. Patients may develop single or multiple lesions, and most lesions persist for seven to fourteen days, healing spontaneously without scarring. Canker sores are neither viral in origin nor contagious. Virus-induced cold sores or fever blisters (herpes simplex) are also common, as is stomatitis, an inflammation often associated with a systemic disease. Periodontal disease, such as gingivitis and Vincent's infection, is likewise seen with some frequency. Candidiasis, a fungal infection with *Candida*, is not uncommon.[49]

Canker sores and sore throat

Although not necessarily related in origin, the same botanicals are used to treat both disorders, so it is convenient to consider them together here. Related ailments, such as cough, were discussed in Chapter 5. Many of these conditions are self-limiting; consequently, herbal treatments are, in general, palliative in character. Some of the commonly used herbs do possess modest antiseptic properties.

Goldenseal

Consisting of the dried rhizome and roots of *Hydrastis canadensis* L., family Ranunculaceae, this native American plant was introduced to the early settlers by the Cherokee Indians. It was long listed in the official compendia (USP and NF) and in relatively recent times was employed in medicine as a bitter. Previously, it had been thought to have value in treating urinary tract infections and in checking internal hemorrhage. However, there is no substantial clinical evidence that goldenseal or its constituents are effective in such conditions.[50]

Goldenseal contains a number of isoquinoline alkaloids, including hydrastine (4 percent), berberine (up to 6 percent), berberastine (3 percent), and canadine (1 percent). Of these, berberine is particularly active, having antibacterial and amoebicidal properties.[51] It probably accounts for the widespread use of goldenseal in the treatment of canker sores and other conditions causing sore mouth. A strong tea prepared from 2 teaspoonfuls of the herb and 1 cup of water has a considerable folkloric reputation as a mouthwash to alleviate pain and facilitate healing. The process may be repeated three or four times daily. Lacking any modern clinical studies dealing with the safety and efficacy of goldenseal when used internally, it is necessary to agree with Sollmann that ingestion of the herb "has few, if any, rational indications."[50]

In recent years, goldenseal has become very popular as a purported immunostimulant. For this purpose, it is commonly combined with echinacea. However, the alkaloids of goldenseal are very poorly absorbed from the gastrointestinal tract, so any significant systemic effects are precluded. In addition, the wild plant has been so intensively collected that the survival of the species is now in doubt. For these reasons, the continued use of this botanical is not rational.

Rhatany

Containing up to 15 percent condensed tannins, the dried root of *Krameria triandra* Ruiz & Pav., family Krameriaceae, is frequently used in the form of a hydroalcoholic solution (tincture) as a treatment for various oral lesions, including inflammation of the gums and of the mucous membranes of the mouth (gingivitis and stomatitis). Used as such or combined with equal parts of myrrh tincture, it is applied locally to noninfectious canker sores with good results. Treatment is carried out two or three times daily. The tincture is also effective when used as a mouthwash, five to ten drops being added to a glass of water for that purpose.[52]

A decoction can also be prepared using 1 level teaspoonful (1.5–2 g) of coarsely powdered root and 5 oz (150 mL) boiling water, covered and kept on the boil for ten to fifteen minutes before being strained. This can be used as a rinse or gargle two to three times daily.

Myrrh

Myrrh is neither plant nor plant part; technically, it is an exudate, an oleo-gum-resin, which exudes from incisions in the bark of *Commiphora mol-mol* Engl., *C. abyssinica* Engl., *C. myrrha* (Nees) Engl., or other species of the same genus (family Burseraceae). The plants yielding myrrh are small trees native to Ethiopia, Somalia, and the Arabian Peninsula.

Consisting of a mixture of about 2.5–10 percent volatile oil, 50–60 percent gum, and 25–40 percent resin composed of resin acids (α-, β-, and γ-commiphoric acids), resenes, and phenolic compounds,[53] the chemical constituents of myrrh are very complex. Presumably, its therapeutic utility may be attributed to the sesquiterpenes that dominate in the essential oil and to the resin acids in the resin. Many of the carbohydrate constituents constituting the gum are insoluble in alcohol and not found in the usual myrrh preparations.[54]

Utilized for many centuries for its astringent and protective properties, myrrh was first listed in the USP in 1820 and enjoyed official status there, and subsequently in the NF, until 1965. It is currently approved by the German Commission E for the local treatment of mild inflammations of the mucous membranes of the mouth and throat. Myrrh is almost always employed in the form of a tincture containing 20 percent of the drug in 85 percent alcohol. The tincture is applied locally to canker sores two or three times daily; a gargle for sore throat consists of five to ten drops of the tincture in a glass of water.[55]

A growing body of evidence suggests that gugulipid, an active component of the guggul or guggulu plant, *Commiphora mukul* Hook., may be effective in lowering serum cholesterol and triglyceride levels. In addition, no significant adverse effects have been reported in clinical studies. Although guggul-based products containing gugulipid are already marketed as drugs in India and France and the powdered resin is available in the form of capsules in the United States, there is insufficient evidence available at this time to render a definitive judgment as to their safety and efficacy.[56] Although preliminary indications are promising, much additional research is required to substantiate the value of the herb. Guggul is therefore not discussed in Chapter 6, "Cardiovascular System Problems." It is mentioned here merely to avoid confusing it with myrrh because both are derived from plants of the same genus.

Sage

The fresh or dried leaves of *Salvia officinalis* L. of the family Lamiaceae have a considerable reputation as a folkloric medicine. One herbal lists more than sixty different conditions for which the plant is said to be useful,[57] but most of the recommendations have not been validated scientifically. Sage leaves contain up to about 2.5 percent of a distinctive-smelling volatile oil in which thujone, the principal constituent (35–60 percent), is accompanied by cineole and other mono- and sesquiterpenes. About 3–7 percent of tannin is also present, as are bitter principles of the diterpene type.[58]

An infusion of sage prepared from 1 teaspoonful (3 g) of finely cut plant material and a cup of boiling water is widely used as a mouthwash or gargle for the treatment of inflammation of the mouth and throat. The volatile oil functions as an antiseptic and the tannin as a local anti-inflammatory agent; the bitter principles produce a pleasant sensory feeling in the mouth and throat.[59] The German Commission E has approved the external use of sage for these purposes, but it also allows its use internally for indigestion and to reduce excessive perspiration.[60] Although the herb may be effective in both conditions, the high thujone content of its volatile oil renders questionable the safety of frequent internal use. Commission E cautions against overuse; in view of the high thujone content of sage oil, it seems wiser to restrict employment of sage preparations to use as mouthwashes or gargles.

Cold sores

Often referred to as fever blisters, cold sores result from an infection with the herpes simplex type 1 virus (HSV-1). This condition is sometimes called herpes simplex labialis because the lesions usually develop on or around the lips, where they tend to recur. HSV-1 is contagious and thought to be transmitted by direct contact. The lesions are self-limiting and heal without scarring, usually within ten to fourteen days. Patients will often associate predisposing factors such as sun or wind exposure, colds and flu, or physical stress and fatigue with the onset of cold sores. Although the lesions are painful and unsightly, treatment is largely symptomatic, utilizing agents that tend to promote healing and, possibly, lessen the pain.[49]

Melissa (balm)

An herb that has shown some promise in the treatment of cold sores is melissa. Consisting of the dried leaves, with or without flowering tops, of *Melissa officinalis* L. (family Lamiaceae), this fragrant herb has been used medicinally for more than twenty centuries. For most of this time, it was employed for its sedative, spasmolytic, and antibacterial properties. These effects are attributed primarily to a volatile oil contained in quality plant material in concentrations of at least 0.05 percent. Some of the chief constituents of the oil include citronellal, citral a, citral b, and many other mono- and sesquiterpenes.[61] The German Commission E has found melissa to be a safe and effective calmative and carminative.[62]

It was first shown in 1978 that an aqueous extract of melissa, containing a variety of polyphenolic substances, including oxidation products of caffeic acid and its derivatives, demonstrated antiviral activity.[63] The caffeic acid oxidation product is said to inhibit not only HSV-1, which causes cold sores, but also the herpes simplex type 2 virus (HSV-2), which causes genital lesions.[64]

A pharmaceutical product for external use is currently marketed in Europe for the treatment of both types 1 and 2 of the herpes simplex virus. It contains a concentrated extract of melissa representing 0.7 g of the leaves per gram of cream-based ointment. Application of the ointment two to four times daily is said to shorten the healing time of the lesions and to decrease their recurrence rate. A similar preparation is available in the United States, or the same can be achieved by preparing a strong tea (infusion) from 2 or 3 teaspoonfuls of the finely cut leaves and 150 mL of water. A pledget of cotton saturated with this solution should be applied to the lesions several times daily.

Much further research is required before the reliability of this remedy for treating cold sores can be assessed. Its efficacy depends on several factors, including the ability of the contained polyphenols to come in contact with the virus. However, untoward side effects of melissa when used for this purpose have not been reported. Further, the treatment is probably as effective as any other self-selected remedy for cold sores.

Dental plaque

Plaque first develops on the enamel of a tooth as a thin pellicle of glycoprotein–mucoprotein derived from saliva. The pellicle harbors acid-producing bacteria that also form long-chain carbohydrate polymers, causing it to thicken into plaque; if it is not removed, calcium salts in the saliva precipitate and convert the plaque into calculus or tartar. Bacteria in the plaque produce acids and other irritants that not only cause tooth decay (caries) but also inflame the gums, causing gingivitis. If untreated, the gingival tissue becomes tender and bleeds, ultimately pulling away from the teeth and leaving pockets that can become infected, a condition known as periodontitis. Periodontal disease is now responsible for 70 percent of the tooth loss in the United States.[49]

Significant herbs

Several herbs are incorporated into antiplaque dentifrices and mouthwashes. They function primarily as antimicrobial agents, inhibiting bacterial adherence to newly formed pellicle and preventing it from being converted into plaque.

Bloodroot (sanguinaria)

One of the most popular of these herbs is bloodroot. The dried rhizome of *Sanguinaria canadensis* L. of the family Papaveraceae contains about 4–7 percent of a mixture of isoquinoline alkaloids, about one-fifth of which is sanguinarine. Bloodroot was once widely used as a stimulating expectorant in various cough preparations; it also enjoyed a considerable folkloric reputation as a treatment for cancer.[65]

Sanguinaria extract, representing a mixture of the total alkaloids, has been incorporated into toothpaste (0.075 percent) and a mouthwash (0.03 percent). A large number of clinical and toxicological studies have been conducted to test the efficacy and safety of these products. The results are mixed, but the consensus is that they may have some merit in plaque and gingivitis reduction.[66] In addition, mouthwashes containing sanguinaria extract are more effective than sanguinaria-containing toothpastes.[67] Effectiveness of sanguinaria applies only to the commercial bloodroot products when used in accordance with the manufacturer's instructions. Use of self-prepared dosage forms of this herb is definitely not recommended.

Minor herbs

Some potentially useful herbs for the prevention of plaque include the leaves of

- neem—*Azadirachta indica* A. Juss.(family Meliaceae);
- mango—*Mangifera indica* L. (family Anacardiaceae);
- basil—*Ocimum tenuiflorum* L. (family Lamiaceae);
- tea—*Camellia sinensis* (L.) O. Kuntze (family Theaceae); and
- curry leaf—*Murraya koenigii* (L.) Spreng. (family Rutaceae).

In India, these individual plant materials are simply rubbed against the teeth to inhibit plaque formation and to treat periodontal disease. Limited in vitro studies confirm the antimicrobial effects of the herbs, and some clinical evidence supports the utility of their aqueous extracts as antiplaque agents in humans. Neem is an ingredient in commercial toothpastes currently marketed in India and Pakistan.[68] Because of the preliminary nature of the clinical studies on the antiplaque value of these five plants, they are not necessarily recommended but, rather, are mentioned here as a matter of record.

References

1. West, D. P., and P. A. Nowakowski. 1996. Dermatitis. In *Handbook of non-prescription drugs*, 11th ed., ed. T. R. Covington, 537–568. Washington, D.C.: American Pharmaceutical Association.
2. Bisset, N. G., ed. 1994. *Herbal drugs and phytopharmaceuticals*, English ed. (*Teedrogen*, M. Wichtl, ed.), 245–247. Boca Raton, FL: CRC Press.
3. *Review of Natural Products:* July 1997.
4. Bisset, N. G., ed. 1994. *Herbal drugs and phytopharmaceuticals*, English ed. (*Teedrogen*, M. Wichtl, ed.), 402–403. Boca Raton, FL: CRC Press.
5. Bisset, N. G., ed. 1994. *Herbal drugs and phytopharmaceuticals*, English ed. (*Teedrogen*, M. Wichtl, ed.), 281–282. Boca Raton, FL: CRC Press.
6. Wormser, H. 1996. Poison ivy, oak, and sumac products. In *Handbook of non-prescription drugs*, 11th ed., ed. T. R. Covington, 647–655. Washington, D.C.: American Pharmaceutical Association.
7. Tyler, V. E. 1985. *Hoosier home remedies*, 124–125. West Lafayette, IN: Purdue University Press.
8. Lipton, R. A. 1958. *Annals of Allergy* 16:526–527.
9. Moore, R. H., and J. D. Bowman. 1996. Burn and sunburn products. In *Handbook of nonprescription drugs*, 11th ed., ed. T. R. Covington, 633–645. Washington, D.C.: American Pharmaceutical Association.

10. Chan, E., and R. Benza. 1996. First-aid products and minor wound care. In *Handbook of nonprescription drugs,* 11th ed., ed. T. R. Covington, 579–599. Washington, D.C.: American Pharmaceutical Association.

11. Smith, E. B. 1998. Fungal diseases of the skin. In *Conn's current therapy,* ed. R. E. Rakel, 823–825. Philadelphia, PA: W. B. Saunders Company.

12. Leung, A. Y., and S. Foster. 1996. *Encyclopedia of common natural ingredients used in food, drugs, and cosmetics,* 25–29. New York: John Wiley & Sons.

13. Agarwala, O. P. 1997. *Drug and Cosmetic Industry* 160 (2): 22–28.

14. Fox, T. R. 1990. *Health Foods Business* 36:45–46.

15. Joshi, S. P. 1998. *Journal of Medicinal and Aromatic Plant Sciences* 20:768–773.

16. Grindlay, D., and T. Reynolds. 1986. *Journal of Ethnopharmacology* 16:117–151.

17. Winters, W. D., R. Benavides, and W. J. Clouse. 1981. *Economic Botany* 35:89–95.

18. Yagi, A., T. Egusa, M. Arase, M. Tanabe, and H. Tsuji. 1997. *Planta Medica* 63:18–21.

19. Klein, A. D., and N. S. Penneys. 1988. *Journal of the American Academy of Dermatology* 18:714–720.

20. McCauley, R. L., J. P. Heggers, and M. C. Robson. 1990. *Postgraduate Medicine* 88 (8): 67–68, 73–77.

21. Fulton, J. E., Jr. 1990. *Journal of Dermatologic Surgery and Oncology* 16:460–467.

22. Visuthikosol, V., B. Chowchuen, Y. Sukwanarat, S. Sriurairatana, and V. Boonpucknavig. 1995. *Journal of the Medical Association of Thailand* 78:403–409.

23. Syed, T. A., S. A. Ahmad, A. H. Holt, S. A. Ahmad, S. H. Ahmad, and M. Afzal. 1996. *Tropical Medicine and International Health* 1:505–509.

24. *Lawrence Review of Natural Products:* April 1992.

25. Newall, C. A., L. A. Anderson, and J. D. Phillipson. 1996. *Herbal medicines: A guide for health-care professionals,* 34–35. London: The Pharmaceutical Press.

26. Schulz, V., R. Hänsel, and V. E. Tyler. 1998. *Rational phytotherapy: A physician's guide to herbal medicine,* 261–262. Berlin: Springer–Verlag.

27. *Bundesanzeiger* (Cologne, Germany): December 5, 1984.

28. *Lawrence Review of Natural Products:* January 1995.

29. Bisset, N. G., ed. 1994. *Herbal drugs and phytopharmaceuticals,* English ed. (*Teedrogen,* M. Wichtl, ed.), 118–120. Boca Raton, FL: CRC Press.

30. *Bundesanzeiger* (Cologne, Germany): March 13, 1986.

31. *Lawrence Review of Natural Products:* October 1995.

32. Weiss, R. F. 1988. *Herbal medicine,* 334–335. Gothenburg, Sweden: AB Arcanum.

33. Gracza, L., H. Koch, and E. Löffler. 1985. *Archiv der Pharmazie* 318:1090–1095.

34. Awang, D. V. C. 1991. *HerbalGram* 25:20–23.

35. Jaarsma, T. A., E. Lohmanns, T. W. J. Gadello, and T. M. Malingré. 1989. *Plant Systematics and Evolution* 167:113–127.

36. *Bundesanzeiger* (Cologne, Germany): June 17, 1992.

37. Kucera, M., M. Barna. O. Horacek, J. Kovarikova, and A. Kucera. 2004. *Wien Medicin Wochenschrade* 152:498–507.

38. Predel, H. G., B. Gianetti, R. Koll, M. Bulitta, and C. Staiger. 2005. *Phytomedicine* 12:707–714.

39. Carson, C. F., and T. V. Riley. 1995. *Journal of Applied Bacteriology* 78:264–269.

40. Babny, P. 1989. *Health Foods Business* 35 (7): 65–66.

41. Penfold, A. R., and F. R. Morrison. 1950. Tea tree oils. In *The essential oils*, vol. 4, ed. E. Guenther, 529–532. New York: D. Van Nostrand.
42. *Lawrence Review of Natural Products:* January 1991.
43. Blackwell, A. L. 1991. *Lancet* 337:300.
44. Bassett, I. B., D. L. Pannowitz, and R. St. C. Barnetson.1990. *Medical Journal of Australia* 153:455–458.
45. Carson, C. F., B. D. Cookson, H. D. Farrelly, and T. V. Riley. 1995. *Journal of Antimicrobial Chemotherapy* 35:421–424.
46. Buck, D. S., D. M. Nidorf, and J. G. Addino. 1994. *Journal of Family Practice* 38:601–605.
47. *Lawrence Review of Natural Products:* May 1995.
48. Hilton, E., H. D. Isenberg, P. Alperstein, K. France, and M. T. Borenstein. 1992. *Annals of Internal Medicine* 116:353–357.
49. Flynn, A. A. 1996. Oral health. In *Handbook of nonprescription drugs*, 11th ed., ed. T. R. Covington, 509–535. Washington, D.C.: American Pharmaceutical Association.
50. Sollmann, T. 1948. *A manual of pharmacology*, 7th ed., 257–258. Philadelphia, PA: W. B. Saunders.
51. Newall, C. A., L. A. Anderson, and J. D. Phillipson. 1996. *Herbal medicines: A guide for health-care professionals*, 151–152. London: The Pharmaceutical Press.
52. Bisset, N. G., ed. 1994. *Herbal drugs and phytopharmaceuticals*, English ed. (*Teedrogen*, M. Wichtl, ed.), 407–408. Boca Raton, FL: CRC Press.
53. Robbers, J. E., M. K. Speedie, and V. E. Tyler. 1996. *Pharmacognosy and pharmacobiotechnology*, 103–104. Baltimore, MD: Williams & Wilkins.
54. Bisset, N. G., ed. 1994. *Herbal drugs and phytopharmaceuticals*, English ed. (*Teedrogen*, M. Wichtl, ed.), 345–347. Boca Raton, FL: CRC Press.
55. *Bundesanzeiger* (Cologne, Germany): August 14, 1987.
56. *Lawrence Review of Natural Products:* February 1995.
57. Keller, M. S. 1978. *Mysterious herbs & roots*, 300–314. Culver City, CA: Peace Press.
58. Bisset, N. G., ed. 1994. *Herbal drugs and phytopharmaceuticals*, English ed. (*Teedrogen*, M. Wichtl, ed.), 440–443. Boca Raton, FL: CRC Press.
59. Hänsel, R. 1991. *Phytopharmaka*, 2nd ed., 95–96. Berlin: Springer–Verlag.
60. *Bundesanzeiger* (Cologne, Germany): May 15, 1985.
61. Bisset, N. G., ed. 1994. *Herbal drugs and phytopharmaceuticals*, English ed. (*Teedrogen*, M. Wichtl, ed.), 329–331. Boca Raton, FL: CRC Press.
62. *Bundesanzeiger* (Cologne, Germany): March 6, 1990.
63. Koch-Heitzmann, I., and W. Schultze. *Zeitschrift für Phytotherapie* 9:77–85.
64. Haas, H. 1991. *Arzneipflanzenkunde*, 148–149. Mannheim, Germany: B. I. Wissenschaftsverlag.
65. *Lawrence Review of Natural Products:* July 1992.
66. Mandel, I. D. 1994. *Journal of the American Dental Association* 125 (suppl. 2): 2S–10S.
67. Grenby, T. H. 1995. *British Dental Journal* 178:254–258.
68. Patel, V. K., and H. Venkatakrishna-Bhatt. 1988. *International Journal of Clinical Pharmacology, Therapy and Toxicology* 26:176–184.

chapter eleven

Performance and immune deficiencies

Performance and endurance enhancers

Stress results when a person is overwhelmed by events and is unable to adapt to them. The condition may be induced by such factors as disease, inappropriate environmental factors, pressures associated with the workplace, inadequate rest, or other physical, chemical, or emotional factors that cause bodily or mental tension. Response to the condition differs for each individual, varying according to the subject's makeup and personality. It may affect mental functions, causing anxiety, depression, fear, guilt, or many other subjective responses. It is well known that, by influencing the autonomic nervous system, stress can also affect bodily functions such as increasing blood pressure, heart rate, sweating, and bowel movements; causing dry mouth and pupillodilation; and altering sleep patterns.

Although the mechanisms are less obvious, stress can trigger or aggravate a wide variety of diseases including diabetes, systemic lupus erythematosus, leukemia, and multiple sclerosis. It is thought that there is a connection between the brain and the immune system via pathways that are as yet unknown. This mind/immune-system connection allows the psychological state of an individual to alter the activity of white blood cells and antibody production. For example, in addition to a physical allergy, hives can be induced by a psychological reaction, and a depressed person is more susceptible to certain infections, such as the common cold.

The treatment and control of stress in an individual may be behavioral, social, psychological, or pharmaceutical. As society becomes more complex and the median age of the general population increases (the elderly are less able to tolerate stress), treatment modalities for stress will become increasingly important.[1]

Basically, two herbal treatments are used to increase resistance to stress: ginsengs and eleuthero. Eleuthero is a relative newcomer to herbal medicine, having been introduced in the Soviet Union in the 1960s. Asian ginseng, on the other hand, has been used continuously in China as a tonic and aphrodisiac for centuries. The use of the word "tonic" in connection with both these herbs has generally been displaced in the herbal literature by the Russian-coined term "adaptogen." This has been defined as an agent that increases resistance to physical, chemical, and biological stress and builds up general vitality, including the physical and mental capacity for work. The Russian term has not been widely adopted in the standard English medical and pharmaceutical literature. It is extensively used in the herbal literature.

Herbal remedies to treat stress

The ginsengs

These consist of the dried root of several species of the genus *Panax* of the family Araliaceae. Asian or Oriental ginseng, extensively cultivated in China, Korea, Russia, and Japan, is *P. ginseng* C. A. Meyer, the most commonly used species. American ginseng is *P. quinquefolius* L. Another species, less frequently encountered, is *sanchi* or *tienchi* ginseng from *P notoginseng* F. H. Chen (Burkill), formerly *P. pseudoginseng* (Burk.) F. H. Chen.[2] Japanese (Chikusetsu) ginseng is *P. japonicus* C. A. Meyer, syn. *P. repens* Maxim. There are qualitative and quantitative differences in the constituents of all three species, but in general, their effects are similar. American ginseng is mostly exported to Asia. Asian ginseng, which is the type that has been most completely investigated, is the ginseng mostly consumed in the West. However, research on American ginseng has been escalating over the past two decades.

The principles believed to be responsible for ginseng's physiological activities are triterpenoid saponin glycosides, generally referred to as ginsenosides but also known, mainly in Japan, as panaxosides. Asian ginseng contains at least thirty ginsenosides, each of which is designated by an uppercase "R" followed by a subscript letter or letter and subscripted numeral or numeral. Its exact composition varies according to the age of the root, the location where grown, the season when harvested, and the method of curing or drying. Because some of the pure ginsenosides produce effects directly opposite to those induced by others and because all are present in the root in relatively small amounts, only the whole root is used in herbal preparations. Asian (Chinese or Korean) ginseng is processed in two ways to produce "white" (root peeled and dried) and "red" (steamed and dried) ginseng (the latter is actually caramel colored).

Hundreds of experiments carried out in small animals have shown that ginseng extracts can prolong swimming time, prevent stress-induced ulcers, stimulate hepatic ribosome production, increase activity of the immune system, stimulate protein biosynthesis, prevent platelet aggregation, and induce many other effects—all of which might contribute to its general tonic or "adaptogenic" effects. Many of these activities have been compared to corticosteroid-like actions, and results of endocrinological studies in animals have suggested that the ginsenosides may augment adrenal steroidogenesis via an indirect action on the pituitary gland by increasing secretion of ACTH (see Figure 8.1, Chapter 8).[5]

In spite of the voluminous literature available on ginseng, the number of reports of reliable human clinical trials is relatively sparse. The results of a total of thirty-seven clinical studies published between 1968 and 1990 have been evaluated. Fifteen of the studies were controlled and eight were double-blind. While taking ginseng preparations, the subjects in thirteen of the studies (1,572 cases) showed improvement in mood; in seventeen studies (846 cases) patients had improved physical performance, and in eleven studies improved intellectual performance was reported. However, the results of the two surveys here summarized were statistically evaluated in only about half of the studies, and they were judged unlikely to conform to current scientific standards regarding study design and conduct.[6] The authors of a more recent review of thirty-five studies (sixteen clinical trials) concluded that their quality was too low to provide convincing evidence of ginseng's effectiveness in improving human physical performance.[7] All the studies reported the absence or near-absence of adverse effects related to ginseng therapy.

More recently, a double-blind study was conducted to study the influence of ginseng on the quality of life in coping with the stress of living in a large city. The 501 male and female volunteers were randomized to take either one capsule containing 40 mg ginseng extract plus multivitamin supplement or the multivitamin supplement alone (control) daily for twelve weeks. The quality of life was assessed monthly by a standardized questionnaire that addressed such issues as perceived well-being, pain, personal satisfaction, depression, energy, sex life, and sleep. The group of subjects taking ginseng extract showed a significant improvement in the quality of life, whereas the control group showed only slight improvement.[8]

Subsequently, in a randomized controlled trial (RCT), 200 mg of the proprietary *P. ginseng* extract Ginsana™ ($n = 15$) or placebo ($n = 15$) was daily administered to healthy, young (eighteen years or older) subjects; aspects of mental health and social functioning, assessed by use of general health status questionnaire, were improved after 4 weeks of therapy, compared to placebo, but the differences were attenuated with continued use.[9] Another study with eighty-three adults (forty women, forty-three men of mean age 25.7 years) with the same extract at two dosages (9,200 and 400 mg) daily found no effect on positive affect, negative affect, or total mood disturbance (all $P > 0.016$).[10]

In the treatment of functional fatigue, a multicenter, comparative, double-blind, clinical study of a total of 232 patients showed that 40 mg of ginseng extract plus multivitamins and minerals taken over forty-two days improved the complaints experienced by patients suffering from fatigue, with tolerability comparable to that of placebo.[11] One should mention, however, that a flaw in the experiment was the lack of vitamins and minerals in the placebo.

Side effects of ginseng consumption attained some notoriety when they were addressed in a 1979 article by Siegel.[12] Because the paper appeared as a "Clinical Note" in the prestigious *Journal of the American Medical Association*, the results of the study were widely disseminated and have been extensively quoted in the herbal literature. Siegel claimed that ingestion of ginseng could result in hypertension, nervousness, irritability, and similar side effects in 14 of 133 volunteers; these effects were grouped under the generic title "ginseng abuse syndrome" (GAS). In a detailed analysis of Siegel's work, Castleman pointed out that there was no control or analysis to determine what types of ginseng were being ingested and, further, some of the subjects were taking excessive amounts—as much as 15 g per day.[13] Also, Blumenthal noted that, apart from serious methodological flaws in the study, it failed to assess the potential influence of accompanying medications, notably caffeine.[14]

Authorities now tend to discount the existence of GAS, but Siegel's study continues to be quoted in the literature. On the basis of its long-term usage and the relative infrequency of reports of significant side effects, it is safe to conclude that ginseng is not usually associated with serious adverse reactions.[5]

In recent years, concern has been expressed about the potential for ginseng affecting the influence of the anticoagulant warfarin (Coumadin®). One case of decreased INR (international normalized ratio) has been reported in which *P. ginseng* was combined with warfarin,[15] apparently paradoxical because ginseng has components with anticoagulant activity; a subsequent study in rats found neither pharmacodynamic nor pharmacokinetic interactions of *P. ginseng* with rats.[16] However, a later RCT found a significant reduction of INR AUC (area under the curve), peak plasma warfarin level, and warfarin AUC following administration of powdered root of *P. quinquefolius* (American ginseng) to twenty young, healthy patients following daily warfarin dosing.[17]

Ginseng is certainly the most costly root; choice specimens with humanoid shapes retail for thousands of dollars. Incidentally, it is this similarity to the human figure that, based on the doctrine of signatures, provides the herb with its ancient reputation as an aphrodisiac. Commercially, ginseng is available in a variety of forms, including teas, capsules, extracts, tablets, roots, chewing gum, cigarettes, and candies. In some of these forms, it is extremely difficult to determine the quality and quantity of the root present. Experiments carried out in the late 1970s showed that 60 percent of fifty-four ginseng products tested were worthless and 25 percent contained no ginseng at all.[18,19] These surveys have been criticized on the basis of inadequate analytical methodology.

Later studies indicated an improvement in compliance of commercial products with indicated species identity but also confirmed wide variation in content of ginsenosides, the presumed main active constituents. Cui and others indicated that forty-four of fifty commercial products from eleven countries contained 1.9–9.0 percent (weigh/weight) of ginsenosides, while six of undetermined identity contained no ginsenosides (one contained ephedrine).[20] Cui found that the ginsenoside content of twenty purported ginseng extract preparations varied from 4.9 to 13.3 percent (w/w).[21]

A later analysis of twenty-five products (purportedly, eight *P. ginseng*, four *P. quinquefolius*, one *P. notoginseng*, nine eleuthero, and three mixtures of various ginsengs) indicated correct labeling but wide variability in concentration of marker compounds. Concentrations of ginsenosides varied by fifteen- to thirty-six-fold in capsules and liquids, respectively, and concentration of eleutherosides varied by forty-three- and two hundred-fold in capsules and liquids, respectively.[22]

The Ginseng Evaluation Program conducted by the American Botanical Council supported the improvement in botanical identification and the continued variation marker content.[23] With a little experience, whole ginseng root is easily recognized organoleptically, so it is recommended that, to ensure quality, the herb be purchased in that form. Alternatively, one must rely on the reputation of the producer or manufacturer.

In summary, ginseng has an ancient reputation as a tonic and aphrodisiac. Some supporting evidence for its effectiveness as an adaptogen has been obtained from small animal and human studies, but additional data from controlled human trials need to be obtained before a definitive evaluation can be made on the therapeutic value of ginseng. The German Commission E has approved ginseng as a tonic to combat fatigue and weakness, as a restorative for declining stamina and impaired concentration, and as an aid to convalescence from illness and degenerative conditions.[24]

Those who choose to consume ginseng for its purported effects do so in the form of a tea prepared from 1 teaspoonful (3 g) of the finely chopped drug and 1 cup (240 mL) of boiling water, covered, and allowed to steep for five to ten minutes before straining. The infusion is taken one to three times a day for three to four weeks.[24] Alternatively, capsules containing 250 mg of the root are used. In short-term use for young and healthy individuals, the dosage is 0.5–1 g of powdered root as two divided doses in the morning and evening on a fasting stomach. Therapy should last for two to three weeks followed by a drug-free period of two weeks between consecutive courses. In long-term use for the elderly and in convalescence, the dose is 0.4–0.8 g daily and can be taken continuously.[5] Ginseng preparations standardized on the basis of their ginsenoside content are available. Dosage is according to the manufacturers' instructions.

Eleuthero

The root of *Eleutherococcus senticosus* (Rupr. & Maxim.) Maxim. of the family Araliaceae was introduced into modern herbal practice some thirty-five years ago by Soviet scientists who were seeking a cheap and abundant substitute for ginseng. Everything possible seems to have been done to make this common plant mimic the scarcer ginseng. When various compounds were isolated from it, they were designated eleutherosides, a name resembling ginsenosides, even though they were not triterpenoid saponins but, rather, were sterols, phenylpropanoid derivatives, lignans, oleanolic acid glycosides, and the like.

When eleuthero was first marketed in the United States in the late 1970s, the name "Siberian ginseng" was utilized, further compounding the confusion.[25] In fact, so many people today confuse eleuthero with ginseng (*Panax* species) that knowledgeable persons almost universally refuse to use the Siberian ginseng designation and commonly refer to the herb by its abbreviated scientific name, eleuthero. A plethora of non-*Panax* species have been inappropriately dubbed "ginseng"—usually qualified by a national association—such as the widely promoted "Brazilian ginseng" (*Pfaffia paniculata*), "Indian ginseng"–Ashwagandha (*Withania somnifera*), and "Peruvian ginseng"–Maca (*Lepidium–meyenii meyenii*), the reputed aphrodisiac.[26]

As noted, eleuthero contains a series of unrelated compounds designated eleutherosides to which the physiologic action—tonic/adaptogenic—has been ascribed. The ginsenosides typical of ginseng are not present.

Eleuthero is often misidentified or adulterated. In 1990, a case in Canada was reported in which a product labeled Siberian ginseng, initially confused with ginseng by the attending physicians, was subsequently determined to be the bark of Chinese silk vine (*Periploca sepium* Bunge of the family Asclepiadaceae), an herb used in traditional Chinese medicine.[27] Another sample of so-called Siberian ginseng that did consist of eleuthero was adulterated by the addition of 0.5 percent caffeine, presumably to provide a stimulant effect to the user.[28] A later case report attributed an elevated digoxin assay to consumption of eleuthoro, where the likely substitution of *Periploca sepium*, which contains cardiac glycosides, probably confounded the digoxin assay.[29,30]

Almost all of the studies dealing with the biological effects of eleuthero have appeared in the Russian literature. According to the summary by Farnsworth and colleagues, studies conducted on a total of twenty-one hundred human subjects with no pathology were designed to determine the ability of humans to withstand various adverse conditions and to improve work output and athletic performance. Male and female subjects from nineteen to seventy-two years of age were given a 33 percent alcoholic extract of eleuthero one to three times daily for periods up to sixty days.[31] The results were generally positive, and no side effects were reported.

Another group of studies involved twenty-two hundred human patients suffering from various pathological conditions, including diabetes, hypertension, cancer, heart disease, and the like. Measurable improvement was said to be effected in many cases.

Eleuthero is widely recommended for, and utilized by, athletes in the United States, who are told that it "stabilizes blood sugar levels during exercise" and "supports the body in adapting to increased levels of mechanical and biochemical stress (ergogenic) that is induced from training and competing."[32] However, this recommendation has been brought into question recently by a well-designed clinical trial with elite athletes (twenty highly trained distance runners). The eleuthero and placebo groups consumed the respective products daily for six weeks and were tested every two weeks for various ergogenic parameters. The conclusion was that ergogenic claims made for eleuthero could not be supported based on trial results.[33]

Some advocates fail to distinguish between ginseng and so-called Siberian ginseng (eleuthero). From the scientific viewpoint, it is apparent that most of the Russian studies of the effects of eleuthero on human beings were not double-blind and also lacked adequate controls. Haas has concluded that the only effect of eleuthero that is adequately documented is that of an immunomodulator.[34] The adaptogen concept, particularly regarding eleuthero, has been challenged as being too vague, and the plant's effects are more properly characterized as antioxidant, anticancer, hypocholesteremic, immunostimulatory, anti-inflammatory, antipyretic, or antibacterial.[35]

Although capsules of eleuthero are readily available on the market, on the basis of existing evidence, its value after evaluating the results as a tonic/adaptogen remains unproven. However, the German Commission E, on the basis of older studies conducted prior to 1991, concluded that eleuthero was an effective tonic.

Sarsaparilla

Consisting of the dried root of various species of the genus *Smilax* that grow in Mexico and Central America, the technical nomenclature of these vining plants is extremely confusing. It is expeditious, therefore, to refer to them by the botanical origins assigned when the drug was last listed in the official compendia in 1965; prior to that time, sarsaparilla had been monographed in the USP and NF for 145 years. Mexican sarsaparilla is obtained from *Smilax aristolochiaefolia* Miller; Honduras sarsaparilla from *S. regelii*, Killip & Morton; Ecuadorian sarsaparilla from *S. febrifuga* Kunth; and Central American sarsaparilla from other undetermined *Smilax* species of the family Smilacaceae.[36]

Originally, the botanical enjoyed an undeserved reputation for the treatment of syphilis. Later it began to be used as a flavoring agent for pharmaceuticals and for soft drinks (root beer). In recent years, it has been extensively advertised in bodybuilding publications as a natural source of testosterone and a legal replacement for illegal androgenic steroids.[37] Some suppliers claim that sarsaparilla actually contains testosterone; others imply that it increases testosterone levels in the body following consumption. Neither assertion is true.

Sarsaparilla does contain several steroids, including sarsapogenin and smilagenin, as well as their glycosides (saponins), which can serve as precursors for the in vitro production of various steroidal drugs. However, these compounds themselves do not function as anabolic steroids and there is no evidence that they or any other sarsaparilla constituents are converted directly to anabolic steroids in the human body. Testosterone itself does not occur in sarsaparilla or in any higher plant.

As if this were not deception enough, it was reported in 1988 that for the ten previous years some commercial herb products labeled sarsaparilla contained instead *Hemidesimus indicus* R. Br., a plant sometimes referred to as false sarsaparilla or Indian sarsaparilla after the country of origin.[38] Belonging to an entirely different plant family (Asclepiadaceae), this substitute contains neither the same saponins nor the other principal constituents found in authentic sarsaparilla. The latter must be obtained from *Smilax* species originating in tropical America.

At present the only legitimate use of sarsaparilla is as a flavoring agent. It has no utility as a performance or endurance enhancer. Because its utility is so often misrepresented and another drug is often substituted for it, the facts concerning sarsaparilla are presented here as a matter of record.

Sassafras

Another herb that cannot be recommended but is still widely used by lay persons as a tonic or performance enhancer and must therefore be mentioned is sassafras. As is the case with sarsaparilla, this native American herb, the root bark of *Sassafras officinalis* Nees & Eberm. (family Lauraceae), was originally recommended as a cure for syphilis. Although sassafras proved ineffective in this regard, it retained its ancient reputation, which became euphemistically expressed as a "blood purifier." In modern parlance, the herb is said to function as a tonic or performance enhancer; in fact, sassafras has been known for more than two centuries to be without significant therapeutic utility.[39]

Sassafras root bark contains 5–9 percent of a highly aromatic oil of which about 80 percent is safrole, a phenolic ether. In 1960, safrole was shown to be carcinogenic in rats and mice, and both sassafras volatile oil and safrole were prohibited by the FDA from use as food additives or flavors.[40] In 1976, bark sold for making sassafras tea was also banned. Nevertheless, the bark is still an article of commerce and is commonly, but unwisely, used as a pleasantly flavored spring tonic. For this reason, the herb is mentioned here, in spite of the fact that it is both unsafe and ineffective.

Ashwagandha

Another herb with purported performance-enhancing effects is ashwa-gandha. The dried root of *Withania somnifera* Dunal (family Solanaceae) is used extensively in Ayurvedic medicine as a tonic. Ayurveda has enjoyed considerable popularity in the United States during recent years, and several herbs used in its traditional healing practices are currently available in tablet form. Ashwagandha is one of these.

Wagner and colleagues have reviewed the evidence supporting the herb's adaptogenic activity.[41] Although preliminary tests in small animals indicate that several contained steroidal derivatives known as sitoindosides may produce some positive effects, substantial clinical evidence obtained from human studies is lacking. Until this is forthcoming, ashwa-gandha cannot be considered a useful adaptogen or tonic.

Cordyceps

A botanical of unusual origin is cordyceps. It is a fungus, *Cordyceps sinensis* (Berk.) Sacc. (family Clavicipitaceae), that parasitizes certain adult caterpillars in the high-mountain regions of China. Historically, this caterpillar fungus has been used in Chinese medicine to aid kidney function, enhance endurance, speed recovery from exhaustion, and promote longevity; however, the rarity of the fungus prevented its widespread use.

Recently, fermentation methods have been developed in China that allow the production of pharmacologically active fungal cells in culture. This fungal cell product has been extensively investigated in China, where it is used for the treatment of chronic bronchitis, kidney disorders, lethargy, and male sexual dysfunction. In the United States, it is marketed as an adaptogen to enhance stamina, reduce symptoms of fatigue, and energize body systems. The active constituents of the fungus have not been identified.[42] In addition, the scientific and medical literature is lacking in clinical studies to support these uses of the drug; consequently, at the present time its effectiveness as an adaptogen remains unproven.

Cancer

Cancer is the second leading cause of death in the United States, producing some five hundred thousand deaths annually, particularly among older people, where it predominates. Cancer (neoplastic) cells are larger and divide more rapidly than normal cells; they serve no useful function. In addition, they metastasize, spreading by various means throughout the body from the primary site. Malignant tumors derived from epithelial tissues are known as carcinomas; those originating from connective, muscle, and bone tissues are called sarcomas.

The origin of cancer (carcinogenesis) has been intensively studied, but definitive results have not been obtained. It apparently results from complicated interactions of viruses, carcinogenic substances or conditions, immunologic factors, and diet. Some of the known carcinogens include radiation (the most dangerous), asbestos, aromatic hydrocarbons and benzopyrene, various alkylating agents, and tobacco. A high-fat diet and excessive use of estrogens have also been implicated.[43]

Because of its life-threatening nature, cancer is not a disease that is amenable to self-treatment. However, some plant drugs are currently used professionally in its treatment with good results, and others are recommended by irresponsible herbalists. It appears worthwhile, therefore, to provide a brief listing and discussion of such products.

Significant anticancer herbs

Catharanthus

The dried whole plant of *Catharanthus roseus* G. Don (family Apocynaceae), formerly designated *Vinca rosea* L., contains more than seventy alkaloids, two of which—vinblastine and vincristine—are extensively used in the treatment of a wide variety of malignant neoplasms.[44] Vinblastine is a mainstay in treating Hodgkin's disease and metastatic testicular carcinoma, and vincristine is used mainly in combination with other anticancer drugs for the treatment of acute lymphocytic leukemia in children, lymphomas, breast cancer, sarcomas, and various childhood neoplasms. Both of these alkaloids are prescription drugs requiring administration by a qualified physician.

Because catharanthus is not employed in herbal form or as a phytomedicine, it is mentioned here to complete the record of plants whose constituents are successfully used in the treatment of cancer. In addition, the successful introduction of these anticancer alkaloids into the clinic in the early 1960s provided great impetus to the research community to investigate plants for novel anticancer agents. Nature creates unexpected and often amazingly novel chemical structures that potentially have new biological activity that can be used therapeutically against cancer. The success of such searches has resulted in the development of podophyllum and Pacific yew as sources of important anticancer agents.

Podophyllum

Consisting of the dried rhizome and roots of *Podophyllum peltatum* L. (family Berberidaceae), podophyllum, or mayapple, contains about 5 percent of an amorphous resin sometimes referred to as podophyllin. Formerly employed as a drastic purgative, podophyllum resin is now used in the form of an alcoholic solution as a topical treatment for certain papillomas (benign epithelial tumors). Etoposide, a semisynthetic derivative of one of the resin constituents, podophyllotoxin, has been developed. It is administered intravenously for the treatment of testicular and germ cell cancers, lymphomas, small-cell lung cancers, refractory acute myelogenous leukemia, refractory acute lymphocytic leukemia in children, Wilm's tumor, neuroblastoma, and AIDS-related Kaposi's sarcoma.[45]

Pacific yew

The bark of the Pacific yew, *Taxus brevifolia* Nutt. (family Taxaceae), contains a small amount (0.01 percent) of taxol or paclitaxel, a compound that has been found to be very useful when administered intravenously for the treatment of advanced ovarian cancer, metastatic breast carcinoma, and AIDS-related Kaposi's sarcoma. A major obstacle in the development of taxol has been the limited supply of the drug. Because of taxol's complex chemistry, it is not economically feasible to synthesize the drug. Supplies initially were limited to small yields of drug extracted from the bark of the old growth of the slowly growing Pacific yew, which was destroyed in the process.

Efforts have been successful in producing the drug in larger yields using semisynthetic methods. The taxol precursor 10-desacetylbaccatin III can be isolated from the needles of *Taxus baccata* L., the common English yew. It is converted to taxol by relatively simple synthetic procedures, and the needles can be harvested without damage to this more rapidly growing plant, making for a renewable resource of the drug.[46]

Unproven anticancer herbs

Phytomedicinals that have no proven value in the treatment of cancer but are nevertheless recommended for its treatment in the uncritical herbal literature are apricot pits, pau d'arco, and mistletoe.

Apricot pits

Much less popular than they were a decade ago, the kernels of *Prunus armeniaca* L. of the family Rosaceae and their contained laetrile (up to 8 percent)—or, more accurately, amygdalin cyanide-containing principles—continue to be used occasionally as a cancer cure. An extensive clinical study begun by the National Cancer Institute in 1980 concluded that laetrile and natural products containing it were "ineffective as a treatment for cancer."[47] Apricot pits and laetrile are mentioned here only to call the reader's attention to their lack of therapeutic value.

Pau d'Arco

The bark of various South and Central American *Tabebuia* species (family Bignoniaceae) is sold under the name of pau d'arco, lapacho, or taheebo and used as a tea to treat various cancers. Presumably, it is purported to be effective because of its content of lapachol derivatives, yet lapachol itself was found to be too toxic for human use in clinical trials.[48,49] The effectiveness of pau d'arco in the treatment of cancer or any other condition remains unproven, and the bark cannot be recommended.

Mistletoe

Because of their reputation as toxic plants, neither European mistletoe, *Viscum album* L., nor American mistletoe, *Phoradendron leucarpum* (Raf.) Rev. & M. C. Johnst., is customarily available from commercial herb supply houses in the United States. Both plants contain similar toxic polypeptides known respectively as viscotoxins and phoratoxins, as well as lectins (glycoproteins) and many other physiologically active constituents, such as tyramine, histamine, and flavonoids.[50]

European mistletoe has an ancient folkloric reputation as a treatment for hypertension, but in recent years much attention has been devoted to its potential antineoplastic effects when aqueous extracts of the plant are administered by subcutaneous injection. Between 1985 and 1990, some sixty articles on this subject were published in the medical literature.[51] The most important active principles responsible for mistletoe's cytotoxic and immunomodulating properties in the treatment of cancer appear to be the lectins.

Unfortunately, because of differences in processing methods and host tree sources, different preparations have different proportions of constituents; consequently, clinical trials using one preparation are not necessarily valid for other preparations. In reviewing the literature, Kleijnen and Knipschild found eleven controlled clinical trials that met predefined criteria for good methodology. None of the studies, however, employed a double-blind design. They found the average quality of the trials poor; the results, based mostly on survival time, were slightly in favor of mistletoe treatment.[52]

At the present time, mistletoe extract and mistletoe lectins are being studied extensively, particularly in Germany, as immunostimulatory adjuncts to use along with standard modalities in the treatment of cancer. For example, one recent study found that in thirty-five patients suffering from malignant stage III/IV glioma, subcutaneous injections of mistletoe extract had an immunostimulatory effect that correlated with an improved quality of life as determined by a standard questionnaire.[53]

Nevertheless, mistletoe's effectiveness as an anticancer treatment is moot in the United States because suitable injectable preparations are unavailable there. In Europe, where they are available, their utility as nonspecific palliative therapy for malignant tumors remains unproven and is controversial. In addition, the German Commission E has declared that the use of the herb to treat hypertension requires substantiation.[54]

Chemopreventive herbs

Chemoprevention can be defined as the use of chemical agents to prevent the occurrence of cancer. Animal studies have demonstrated that several hundred different compounds have shown chemopreventive properties, and the consequence of this large number of compounds is that they probably act through a variety of different mechanisms. Epidemiological data on cancer indicate that chemopreventive agents in the diet may play an important role in preventing cancer. It is known that vegetarian populations have significantly lower incidences of breast, colon, and prostate cancer. Whereas the cause of this lower incidence may be related to the decreased amount of fat in the vegetarian diet, another factor is the abundance of chemopreventive agents in vegetarian diets, including terpenes, organosulfides, isothiocyanates, indoles, dithiolethiones, polyphenols, flavones, tannins, carotenoids, and vitamins.[55,56]

Specific mechanisms for chemoprevention are varied. Some dietary agents may block carcinogen activation by inducing phase II enzymes involved in precarcinogen xenobiotic detoxification,[55] leading to a more rapid excretion from the body. Other dietary compounds—for example, carotenes, vitamins C and E, and some flavonoids—have antioxidant activity that prevents the formation of free radicals (see Figure 6.3, Chapter 6). Free radicals can induce mitochondrial and nuclear DNA damage leading to carcinogenesis. Several herbal products appear to have potential as effective chemopreventive agents.[57]

Green tea

The leaves and leaf buds of *Camellia sinensis* (L.) Kuntze (family Theaceae), an evergreen shrub indigenous to mountainous regions of Southeast Asia, are carefully processed in China and Japan to produce green tea. It is prepared by rapidly drying the freshly harvested leaves in copper pans over a mild artificial heat or by first subjecting the fresh leaves to steam for at least thirty seconds and then drying. Both processes inactivate oxidative enzymes (primarily polyphenol oxidase), which prevent the destruction of polyphenols (catechins), the antioxidant constituents of tea. Fermentation of the withered, rolled, and crushed fresh leaves at high humidity produces black tea, which contains the less active oxidation and condensation products of catechins.[58]

The chemopreventive constituents of green tea are the polyphenols, which are complex flavonoid structures also called catechins. The main polyphenols are epicatechin, epicatechin gallate, epigallocatechin gallate

(EGCG), and proanthocyanidins. Because they are water soluble, a cup of green tea contains 80–140 mg of polyphenols as well as 50 mg of caffeine. Commercially available green tea extracts are standardized to 60–80 percent polyphenols.

Green tea and green tea polyphenols have been shown to demonstrate chemopreventive and anticancer effects in a number of animal tumor models. The polyphenols were found to scavenge hydrogen peroxide and superoxide anions (Figure 6.3, Chapter 6) and reduce oxygen free-radical damage. In addition, they increase antioxidant activity and phase II enzymes in mouse skin, liver, lungs, and small bowel. Green tea polyphenols—particularly EGCG—were more potent scavengers of polycyclic aromatic hydrocarbon free radicals (carcinogens) than the widely used food antioxidant BHT.[59]

Although epidemiological studies and clinical trials indicate a potential benefit of green tea consumption in the prevention of cancer, a direct cause-and-effect relationship has not been proved through clinical trials in humans. Because of the lack of side effects and even though the chemopreventive action of green tea still must be determined, the use of green tea or green tea extract at dosages normally consumed by humans may have beneficial effects. The chemopreventive daily dose of green tea has not been established. Overconsumption may produce effects associated with excessive caffeine intake, such as insomnia, nervousness, and tachycardia.

Grape seed extract

The seeds of the grape, *Vitis vinifera* L. (family Vitaceae), and its varieties contain polyphenolic proanthocyanidins, which are oligomers or polymers of catechin and epicatechin. In turn, the proanthocyanidins can bind to each other to form oligomers known as procyanidin, oligomeric proanthocyanidin complexes (OPCs), procyanidolic oligomers (PCOs), leucoanthocyanins, condensed tannins, or pycnogenol.[60] Commercial preparations of grape seed extract are usually standardized on the basis of procyanidin content.

Procyanidins are strong antioxidants, scavengers of free radicals, and inhibitors of lipid peroxidation. Grape seed extract, vitamin C, and vitamin E were compared using two in vitro assays of free-radical quenching ability. Grape seed extract was found to be a more efficient antioxidant in the assays than the antioxidant vitamins C and E.[61] Because of these properties, grape seed extract has considerable potential as a chemopreventive agent. However, clinical trials in humans are lacking to support its use for this purpose. A few studies have been reported supporting its use to treat certain circulatory disorders such as venous insufficiency and retinopathy. It has a low toxicity profile and side effects have not been reported. A chemopreventive dose has not been established, but a daily dose of 50 mg has been recommended for general health purposes.[62]

Some confusion exists concerning the term pycnogenol. In the United States, it is a registered trademark for the procyanidins extracted from the bark of the French maritime pine, *Pinus pinaster* Ait., family Pinaceae. Pycnogenol contains slightly fewer procyanidins than grape seed extracts; other than that, both products appear to be equivalent.[63]

Communicable diseases and infections

One way to combat communicable diseases and infections is to stimulate the body's own immune system to resist the unwanted microorganisms. In general, this is called immunomodulation. Specifically, when an augmented or enhanced immune response is required, it is known as biological response modification. The mechanism of immunomodulation is complex, and a rather daunting nomenclature has grown up around the various components and factors involved.

The immune system can be separated into two functional divisions. One is innate immunity, which involves the stimulation of cells that nonspecifically recognize foreign invaders and destroy them. The other is adaptive immunity, which differs from the innate immune response in two critical areas: specificity and memory. A clear distinction of the adaptive division from the innate division is not possible because both divisions heavily interact.

The major effector cells of the innate system are illustrated in Figure 11.1. They include large granular lymphocytes, also called natural killer (NK) cells. They are ready to kill a variety of target cells such as tumor cells or viral-infected host cells as soon as they are formed rather than requiring the maturation and activation process required by the cells of the adaptive immunity process. Polymorphonuclear leukocytes constitute the primary human defense against pathogenic bacteria and migrate to sites of infection in a process termed chemotaxis. These effector cells are capable of phagocytizing and killing microorganisms without help; however, they function more efficiently when pathogens are first opsonized (coated) by components of the complement system or by antibodies.

Macrophages and monocytes are mononuclear cells of the innate immune system capable of phagocytosis. They also have the ability to release soluble factors with inflammatory properties. Monocytes are found circulating in the bloodstream, and the macrophages are fixed to the tissue of the reticuloendothelial system, located in the liver, spleen, GI tract, lymph nodes, brain, and others. Both monocytes and macrophages act as antigen-presenting cells to stimulate the adaptive immune system. Phagocytosis by these cells results in the digestion of the infecting organism into small peptide fragments that are antigenic. The antigenic fragments are combined with major histocompatibility complex (MHC) proteins on the surface of the antigen-presenting cell, where they are

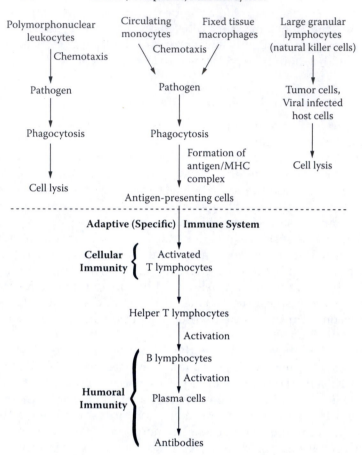

Figure 11.1 Schematic model of intracellular interactions between effector cells of the immune system.

recognized by the T cell receptors on the surface of T lymphocytes. MHC constitutes a group of molecules important in helping the body distinguish self from nonself.

The main types of effector cells of the adaptive immune system are T lymphocytes (T cells), which protect against intracellular diseases and cellular neoplasms, and B lymphocytes, which develop into plasma cells that secrete antibodies or immunoglobulins in response to certain antigens. The T lymphocytes are classified into three types, depending on their function: T helper cells secrete lymphokines that stimulate the activity of T killer cells, T suppressor cells control numerous immune reactions, and T killer cells produce tumor necrosis factor.

In summary, T lymphocytes produce cellular immune responses, and plasma cells derived from B lymphocytes, by the production of antibodies, induce humoral immune responses. Macrophages secrete so-called cytokines that influence all immune responses; lymphocytes communicate through the secretion of lymphokines.[64]

Biological response modifiers (BRMs) or immunostimulants may affect the cellular or humoral immune system or both. They are nonspecific in character, producing general stimulation of the entire system. Because the response capacity of the immune system is limited, BRMs are more effective when used in conjunction with other chemotherapeutic agents and when the disease entity is quantitatively small.

Herbal remedies for communicable diseases and infections

Echinacea

Of all the nonspecific immunostimulants of plant origin, the most comprehensively studied is echinacea. Of the established nine species (and four varieties)[65] of indigenous North American herbaceous perennials, only three are purported components of commercial medicinal plant preparations: *Echinacea angustifolia* DC., the narrow-leafed purple coneflower; *E. pallida* (Nutt.) Nutt., the pale purple coneflower; and *E. purpurea* (L.) Moench, the purple coneflower. All three of these plants are members of the Asteraceae and in today's usage are probably more or less interchangeable.

E. pallida and *E. angustifolia* were the species long recognized in the *National Formulary*; presumably both were utilized in the popular echinacea preparations ("Specific Medicines") once marketed by the Lloyd Brothers of Cincinnati. At the present time, Foster advocates the use of *E. purpurea*, simply because it is the only species presently cultivated. The numerous studies of the herb carried out in Germany have utilized primarily the aboveground parts rather than the rhizome and roots of *E. purpurea*.[66] (It should be noted that a recent Canadian taxonomic revision of the *Echinacea* genus designates *E. angustifolia* and *E. pallida* as *varieties* of *E. pallida*—namely, *E. pallida* var. *angustifolia* and *E. pallida* var. *pallida*.[67])

Endemic or rare echinacea species, including *E. atrorubens*, *E. paradoxa*, and *E. simulate*, have been documented in commercial plant supplies, vicariously harvested or intentionally substituted.[66] Other plants have often been fraudulently substituted for echinacea. At the turn of the century, *Eryngium praealtum* A. Gray, known as button snakeroot, was commonly employed. In recent decades, *Parthenium integrifolium* L., often called Missouri snakeroot, has been a particularly troublesome substitute. *E. angustifolia* has also been mixed with *E. pallida* and marketed as Kansas snakeroot. This tendency toward falsification is especially important to recognize because echinacea is generally marketed as a tincture (hydroalcoholic extract); in this form, substitution is difficult to detect. In the absence of any standards of quality for phytomedicinals, echinacea preparations should be obtained only from suppliers with impeccable reputations for integrity.

Echinacea has no direct bactericidal or bacteriostatic properties. Its beneficial effects in the treatment of infections are caused by its ability to act as an immunostimulant. It stimulates the innate (nonspecific) immune system, resulting in an increase in phagocytosis and effector cell activity. Macrophage and T lymphocyte release of tumor necrosis factors (poly-peptide inflammatory mediators with antitumor activity) and interferons (proteins with antiviral and antitumor activity) is increased. In addition, the spread of infectious agents in the body may be lessened through an inhibition of tissue hyaluronidase. All of these actions tend to increase the body's resistance to bacterial and viral infection.[68]

The exact identity of the principles in echinacea responsible for these effects is still undergoing intensive study. Two high-molecular-weight poly-saccharides with immunostimulatory activity have been isolated from the aerial plant parts of *E. purpurea*. One is a 4-*O*-methylglucuronoarabinoxylan with a molecular weight of 35 kD, and the other is an acid rhamnoarabi-nogalactan with a molecular weight of 50 kD. The main active component in the squeezed juice of flowers of *E. purpurea* is apparently an acidic ara-binogalactan with a molecular weight of 75 kD, which can also be pro-duced in large-scale tissue cultures. This latter polysaccharide increases phagocytosis and the release of interferon and tumor necrosis factor.

The immunostimulant activities of the polysaccharides have only been demonstrated in vitro or following parenteral administration. Whether or not they would be active upon oral administration has not been clari-fied.[69] Stimulation of phagocytosis is apparently also enhanced by com-ponents of the alkamide fraction (mainly isobutylamides) and by caffeic acid derivatives (e.g., cichoric and caftaric acids).[70] Other potential candi-dates as active principles in echinacea are glycoproteins, ketoalkenes, and ketoalkynes (these last often are mistakenly termed "polyacetylenes").[71]

An enormous literature on echinacea currently exists. With the vast amount of data available, as well as a great deal of information and mis-information in the popular herbal literature, it is difficult to summarize briefly and accurately the medical applications of the herb. Although the injectable and the ointment forms are utilized in Europe, neither is gener-ally available in the United States. Therefore, these remarks will be lim-ited to use of the hydroalcoholic extract and the pressed juice, which are generally consumed orally but may also be applied locally.

Researchers in Germany reviewed a total of twenty-six controlled clin-ical trials conducted before 1994 that investigated the immunomodulatory activity of echinacea preparations. Of the total of thirty-four test treat-ment groups, twenty-two were deemed by the reviewers to have given results indicating that echinacea was efficacious as an immunomodulator, principally in upper respiratory infections. The review emphasized, however, that echinacea preparations can have different amounts or content of active constituents, depending on the plant material used and the type of extraction employed. Therefore, an extrapolation of results from one preparation to another is not possible without proof of chemical and pharmaceutical equivalence.[72]

A more recent review, conducted through 2005, assessed sixteen studies: five from the United States, five from Germany, three from Canada, two from Sweden, and one from Russia.[73] The main findings of this systematic review included:

- The variety of commercial available echinacea-containing products assessed contained different amounts of bioactive compounds and therefore could not be considered biochemically comparable.
- The methods used to assess cold variables were highly variable.
- Most of the trials reviewed had reasonable to good methodology, based on assessment of two independent reviewers using the Jadad method of evaluation.[74]
- Preparations based on the aerial parts of *E. purpurea* may be effective in the early treatment of colds in adults, but results are not fully consistent, and there is no clear evidence that other preparations work or that children benefit.
- Adverse side effects associated with echinacea preparations were infrequent or minor and mostly similar to placebo.

A RCT of a "standardized" proprietary *E. purpurea* extract (Echinilin®, Natural Factors Nutritional Products, Inc., Vancouver, BC), described as combined aqueous and alcoholic extracts "of various parts of freshly harvested ... plants," was conducted with 282 subjects, aged eighteen to sixty-five years.[75] Early intervention with this formulation was found to reduce the severity and duration of symptoms in subjects with naturally acquired upper respiratory tract infections (URTIs).

However, this study suffers from a limitation shared by numerous trials of proprietary preparations, in that it is impossible to determine the precise character of the tested formulation. As a result, the best judgment one can make of this treatment is that it has appreciable potential for benefit, as long as manufacturing procedures are sufficiently standardized to ensure a chemically consistent product dependent also on reliable genetic and cultivational factors. "Standardization" to more or less arbitrary levels and ratios of alkamides, cichoric acid, and polysaccharides will not ensure reproducibility of effect of an echinacea preparation.

Oral consumption seems to be utilized primarily, but not exclusively, for preventing and treating the common cold and its associated conditions, such as sore throat. The German Commission E has approved alcoholic extracts of the root of *E. pallida* for the supportive treatment of flu-like infections and pressed juice from the aerial parts of *E. purpurea* for the supportive treatment of recurrent infections of the upper respiratory tract and lower urinary tract. The words "supportive treatment" denote that echinacea would ordinarily be administered together with other antibacterial agents and antibiotics.

The commission also has approved its local application for the treatment of hard-to-heal superficial wounds.[76] Preparations of the underground parts of other *Echinacea* spp. apparently are similarly useful in all of these conditions. Echinacea's value in the treatment of other conditions, including yeast infections, side effects of radiation therapy, rheumatoid arthritis, cancer, etc.—although often extolled in modern herbals intended for popular consumption—remains unproven.[77]

Dosage of echinacea is dependent on the potency of the particular sample or preparation utilized. The producer of a typical American hydroalcoholic preparation recommends fifteen to thirty drops (0.75–1.5 mL) two to five times daily. The German Commission E recommends a daily dose of root fluidextract (20 percent) equivalent to 900 mg of crude drug, and 6–9 mg is recommended for the pressed juice or equivalent amounts of preparations made from it. In the United States, the former official dose of the rhizome and roots was 1 g/day. Capsules containing it in powdered form are available. A decoction prepared by simmering 2 teaspoonfuls of the coarsely powdered herb in a cup of boiling water is sometimes used but is not recommended because not all of the active constituents are water soluble.[78]

In a personal communication to C. Hobbs, Bauer has indicated that some persons in Germany believe echinacea works by stimulating lymphatic tissue in the mouth, thereby initiating an immune response.[71] Although evidence in support of this assertion is scanty, if it is true, it would support the use of a hydroalcoholic solution as the preferred dosage form. It would also suggest that such preparations would be more effective if held in the mouth for a period of time prior to swallowing.

Due to the potential for stimulating autoimmune processes, echinacea should not be used by persons suffering from severe systemic illnesses such as tuberculosis, leukosis, collagen diseases, and multiple sclerosis. Commission E recommends that neither internal nor external use should exceed a period of eight successive weeks. Infrequent allergies may occur, especially in patients allergic to plants of the sunflower family (Asteraceae).

Cat's claw

Cat's claw consists of the dried root bark of *Uncaria tomentosa* (Willd.) DC. (family Rubiaceae), a woody vine growing in South America. Cat's claw refers to the little curved-back spines on the stem at the leaf juncture.

Aqueous extracts of the root bark have been used in Peruvian traditional medicine as an anti-inflammatory, contraceptive, and cytostatic remedy.[79] A related species, *U. guianensis* (Aubl.) Gmel., has also had a long folkloric use in South America as a wound healer and for treating intestinal ailments.[80]

It has been demonstrated that cat's claw contains pentacyclic oxindole alkaloids that have immunostimulant properties.[81] However, it is of interest that two chemical types of *U. tomentosa* that have different alkaloid patterns occur in nature. In one chemotype, the plant contains the pentacyclic oxindole alkaloids; the other contains tetracyclic oxindoles that act on the central nervous system and exert antagonistic effects on the action of the pentacyclic oxindoles. Mixtures of these two chemotypes would therefore be unsuitable for immunomodulation therapy.[82]

Although there is little published data on the toxicology of cat's claw, the limited number of pharmacological studies seems to indicate little hazard in ingesting the plant decoction. At the same time, human clinical trials supporting its immunomodulatory efficacy are lacking, so at this time the drug cannot be recommended.

Andrographis

Extracts of the leaves of *Andrographis paniculata* (Burm. f.) Nees have been used for both prophylactic and symptomatic treatment of URTIs such as the common cold, as well as for uncomplicated sinusitis, pharyngotonsilitis, pneumonia, and bronchitis.[83] The diterpenic lactone andrographolide and its derivatives—namely, 14-deoxyandrographolide, 14-deoxy-11,12-didehydroandrographolide, and neoandrographolide—are regarded as the herb's active constituents, with immunostimulant,[84] antipyretic, anti-inflammatory,[85] and antidiarrheal properties.[86] Extracts of the herb are normally standardized to a content of 4–6 percent total andrographolides.

A systematic review of RCTs for symptomatic treatment of uncomplicated URTIs was conducted in 2004.[87] Only studies that provided data concerning the assessment of symptom severity were deemed eligible for review. Four studies meeting the eligibility criteria were identified. Two trials ($n = 225$) evaluated the efficacy of an andrographis leaf extract alone or in fixed combination with an eleuthero root extract versus placebo. In both trials, andrographis preparations were judged superior to placebo in reducing the severity of symptoms associated with URTIs.

Another trial evaluated the efficacy of andrographis versus placebo in 208 patients with the common cold; fewer days of sick leave were registered by subjects administered andrographis, and total recovery as well as symptom relief was greater in the treatment group than in the placebo group. In the fourth trial, the efficacies of andrographis and paracetamol were compared in 152 patients suffering from pharyngotonsilitis; after three days, patients administered 6 g of dried andrographis leaves daily experienced fever loss and sore throat eradication at levels comparable to those taking 3.9 g of paracetamol daily.

Two randomized parallel-group clinical studies conducted in Volgograd tested a proprietary fixed combination of extracts of andrographis leaves and eleuthero root (Kan Jang®, Swedish Herbal Institute, Göteborg, Sweden) in the treatment of influenza.[88] Patients received either two or three Kan Jang tablets (each containing 88.8 mg standardized andrographis extract and 10.0 mg standardized eleuthero extract) three times daily or conventional treatment. In the first trial, a pilot study involving 540 patients, seventy-one were treated with Kan Jang; in the second, sixty-six patients received the herbal treatment. The differences in duration of sick leave and frequency of postinfluenza complications indicated that the Kan Jang phytopreparation not only contributed to quicker recovery but also reduced the risk of postinfluenza complications. Kan Jang was also well tolerated by patients.

The contribution of eleuthero extract to efficacy of Kan Jang has not been determined, and the special andrographis extract (SHA-10) alone has been found effective in alleviating the symptoms of the common cold in an RCT.[89]

Kan Jang is not to be confused with Kan Jang™ oral solution (also from the Swedish Herbal Institute), a fixed combination of standardized extracts of *Echinacea purpurea*, *Adhotada vasica*, and *Eleutherococcus senticosus* in a "liquid matrix ... containing liquorice, nipagin, nipasol, sorbitol, polysorbate, eucalyptus oil, peppermint oil, coltsfoot leaf aroma, ginger extract and water."[90]

References

1. Berkow, R., ed. 1997. Psychosomatic disorders. In *The Merck manual of medical information*, home ed., 390–392. Whitehouse Station, NJ: Merck Research Laboratories.
2. Wen, J. 2001. Utilization of biotechnological, genetic and cultural approaches for North American and Asian ginseng improvement. *Proceedings of the International Ginseng Workshop*, ed. Z. K. Punja, 67–88. Vancouver, Canada: Simon Fraser University Press.
3. Court, W. E. 2000. *Ginseng: The genus* Panax, 226. Amsterdam: Harwood Academic Publishers.
4. Sticher, O. 1998. *Chemtech* April: 26–32.
5. Newall, C. A., L. A. Anderson, and J. D. Phillipson. 1996. *Herbal medicines: A guide for health-care professionals*, 145–150. London: The Pharmaceutical Press.
6. Schulz, V., R. Hänsel, and V. E. Tyler. 1998. *Rational phytotherapy*, 3rd ed., 270–272. Berlin: Springer–Verlag.
7. Bahrke, M. S., and W. P. Morgan. 2000. *Sports Medicine* 29 (2): 113–133.
8. Caso Marasco, A., R. Vargas Ruiz, A. Salas Villagomex, and C. Begona Infante. 1996. *Drugs under Experimental and Clinical Research* 22:323–329.
9. Ellis, J. M., and P. Reddy. *The Annals of Pharmacotherapy* 36:375–379.

10. Cardinal, B. J., and H.-J. Engles. 2001. *Journal of the American Dietetic Association* 101 (6): 655–660.
11. LeGal, M., P. Cathebras, and K. Strüby. 1996. *Phytotherapy Research* 10:49–53.
12. Siegel, R. K. 1979. *Journal of the American Medical Association* 241:1614–1615.
13. Castleman, M. 1990. *The Herb Quarterly* 48:17–24.
14. Blumenthal, M. 1991. *Whole Foods* March: 89–93.
15. Janetzky, K., and A. P. Morreale. 1997. *American Journal of Health-Systems Pharmacy* 54:692–693.
16. Zhu, M., K. W. Chan, L. S. Ng, Q. Chang, S. Chang, and R. C. Li. 1999. *Journal of Pharmacy and Pharmacology* 51:175–180.
17. Yuan, C.-S., G. Wei, L. Dey, T. Karrison, L. Nahlik, S. Maleckar, K. Kasza, M. Ang-Lee, and J. Moss. 2004. *Annals of Internal Medicine* 141 (1): 23–27.
18. Liberti, L. E., and A. Der Marderosian. 1978. *Journal of Pharmaceutical Sciences* 67:1487–1489.
19. Ziglar, W. 1979. *Whole Foods* 2 (4): 48–53.
20. Cui, J., M. Garle, P. Eneroth, and I. Biökhem. 1994. *Lancet* 344:134.
21. Cui, J. 1995. *European Journal of Pharmaceutical Sciences* 3:77–85.
22. Harkey, M. R., G. L. Henderson, M. E. Gershwin, J. S. Stern, and R. M. Hackman. 2001. *American Journal of Clinical Nutrition* 73:1101–1106.
23. Ginseng Evaluation Program, Austin, TX, American Botanical Council, 2001.
24. Bisset, N. G., ed. 1994. *Herbal drugs and phytopharmaceuticals*, English ed. (*Teedrogen*, M. Wichtl, ed.), 236–238. Boca Raton, FL: CRC Press.
25. Foster, S. 1991. Siberian ginseng *Eleutherococcus senticosus*, Botanical series no. 302, American Botanical Council, Austin, TX, 7 pp.
26. Awang, D. V. C. 2003. *HerbalGram* 57:31–35.
27. Awang, D. V. C. 1991. *Journal of the American Medical Association* 266:363.
28. Awang, D. V. C. Personal communication, May 21, 1991.
29. McRae, S. 1996. *Canadian Medical Association Journal* 155 (3): 293–295.
30. Awang, D. V. C. 1996. *Canadian Medical Association Journal* 155 (9): 1237.
31. Farnsworth N. R., A. D. Kinghorn, D. D. Soejarto, and D. P. Waller. 1985. Siberian ginseng (*Eleutherococcus senticosus*): Current status as an adaptogen. In *Economic and medicinal plant research*, vol. 1, ed. H. Wagner, H. Hikino, and N. R. Farnsworth, 155–215. Orlando, FL: Academic Press.
32. Reaves, W. 1991. *Health Foods Business* 37 (3): 56–60.
33. Dowling, E. A., D. R. Redondo, J. D. Branch, S. Jones, G. McNabb, and M. H. Williams. 1996. *Medicine and Science in Sports and Exercise* 28:482–489.
34. Haas, H. 1991. *Arzneipflanzenkunde*, 135–136. Mannheim, Germany: B. I. Wissenschaftsverlag.
35. Davydov, M., and A. D. Krikorian. 2000. *Journal of Ethnopharmacology* 72:345–393.
36. Tyler, V. E., L. R. Brady, and J. E. Robbers. 1988. *Pharmacognosy*, 9th ed., 486. Philadelphia, PA: Lea & Febiger.
37. Tyler, V. E. 1988. *Nutrition Forum* 5:23.
38. Blumenthal, M. 1988. *Health Foods Business* 34 (4): 58.
39. Tyler, V. E. 1987. Some potentially useful drugs identified in a study of Indiana folk medicine. In *Folklore and folk medicines*, ed. J. Scarborough, 98–109. Madison, WI: American Institute of the History of Pharmacy.
40. Crellin, J. K., and J. Philpott. 1990. *Herbal medicine past and present*, vol. 2, 25. Durham, NC: Duke University Press.

41. Wagner, H., H. Nörr, and H. Winterhoff. 1992. *Zeitschrift für Phytotherapie* 13:42–54.
42. Walker, L. A. 1997. *Drug Topics, Natural Products Update Supplement* June: 11–12.
43. Balmer, C., and A. W. Valley. 1997. Basic principles of cancer treatment and cancer chemotherapy. In *Pharmacotherapy: A pathophysiologic approach,* 3rd ed., ed. J. T. DiPiro, R. L. Talbert, G. C. Yee, G. R. Matzke, B. H Wells, and L. M. Posey, 2403–2465. Stamford, CT: Appleton & Lange.
44. Robbers, J. E., M. K. Speedie, and V. E. Tyler. 1996. *Pharmacognosy and pharmacobiotechnology,* 169–171. Baltimore, MD: Williams & Wilkins.
45. Robbers, J. E., M. K. Speedie, and V. E. Tyler. 1996. *Pharmacognosy and pharmacobiotechnology,* 137–138. Baltimore, MD: Williams & Wilkins.
46. Buss, A. D., and R. D. Waigh. 1995. Natural products as leads for new pharmaceuticals. In *Burger's medicinal chemistry and drug discovery,* 5th ed., vol. 1, ed. M. E Wolf, 983–1033. New York: John Wiley & Sons, Inc.
47. Moertel, C. G., T. R. Fleming, J. Rubin, L. K. Kvols, G. Sarna, R. Koch, V. E. Currie, C. W. Young, S. E. Jones, and J. P. Davignon. 1982. *New England Journal of Medicine* 306:210–206.
48. *Lawrence Review of Natural Products:* July 1990.
49. Owsald, E. H. 1993/1994. *British Journal of Phytotherapy* 3:112–117.
50. Newall, C. A., L. A. Anderson, and J. D. Phillipson. 1996. *Herbal medicines: A guide for health-care professionals,* 193–196. London: The Pharmaceutical Press.
51. Bowman, I. A. 1990. *Texas Heart Institute Journal* 17:310–314.
52. Kleijnen, J., and P. Knipschild. 1994. *Phytomedicine* 1:255–260.
53. Lenartz, D., B. Stoffel, J. Menzel, and J. Beuth. 1996. *Anticancer Research* 16:3799–3802.
54. *Bundesanzeiger* (Cologne, Germany): December 5, 1984.
55. Wattenberg, L. W. 1993. *Cancer Research* 53:5890–5896.
56. Stavric, B. 1994. *Clinical Biochemistry* 27:319–332.
57. Troll, W., J. S. Lim, and K. Frenkel. 1994. Prevention of cancer by agents that suppress production of oxidants. In *Food phytochemicals for cancer prevention II: Teas, spices, and herbs,* ed. C.-T. Ho, T. Iosawa, M.-T. Huang, and R. T. Rosen, 116–121. Washington, D.C.: American Chemical Society.
58. Gutman, R. L., and B.-H. Ryu. 1996. *HerbalGram* 37:33–45.
59. Snow, J. 1995. *The Protocol Journal of Botanical Medicine* Autumn: 47–51.
60. Bombardelli, E., and P. Morazzoni. 1995. *Fitoterapia* 66:291–317.
61. Bagchi, D., A. Garg, R. L. Krohn, M. Bagchi, M. X. Tran, and S. J. Stohs. 1997. *Research Communications in Molecular Pathology and Pharmacology* 95:179–189.
62. Murray, M. T. 1995. *The healing power of herbs,* 2nd ed., 184–191. Rocklin, CA: Prima Publishing.
63. *Lawrence Review of Natural Products:* February 1991.
64. Hall, P. D., and J. A. Tami. 1997. Function and evaluation of the immune system. In *Pharmacotherapy: A pathophysiologic approach,* 3rd ed., ed. J. T. DiPiro, R. L. Talbert, G. C. Yee, G. R. Matzke, B. H Wells, and L. M. Posey, 1600–1647. Stamford, CT: Appleton & Lange.
65. McGregor, R. L. 1996. *University of Kansas Science Bulletin* 48:113–142.
66. Foster, S. 1991. *Echinacea: Nature's immune enhancer.* Rochester, VT: Healing Arts Press, 150 pp.
67. Binns, S. E., B. R. Baun, and J. T. Arnason. 2002. *Systematic Botany* 27:610–632.

68. Newall, C. A., L. A. Anderson, and J. D. Phillipson. 1996. *Herbal medicines: A guide for health-care professionals,* 101–103. London: The Pharmaceutical Press.
69. Bauer, R., and H. Wagner. 1991. *Echinacea* species as potential immunostimulatory drugs. In *Economic and medicinal plant research,* vol. 5, ed. H. Wagner and N. R. Farnsworth, 253–321. London: Academic Press.
70. Bauer, R., P. Remiger, K. Jurcic, and H. Wagner. 1989. *Zeitschrift für Phytotherapie* 10:43–48.
71. Bauer, R. 2000. Chemistry, pharmacology and clinical applications of echinacea products. In *Herbs botanicals and teas,* ed. G. Mazza and B. D. Oomah, 45–73. Basel: Technomic Publishing Co., Inc.
72. Melchart, D., K. Linde, F. Worku, R. Bauer, and H. Wagner. 1994. *Phytomedicine* 1:245–254.
73. Linde, K., B. Barrett, K. Wölkart, R. Bauer, and D. Melchart. 2006. *The Cochrane Database of Systematic Reviews* 1:1–39, article no. CD 000530. PUB 2.
74. Jadad, A. R., R. A. Moore, D. Carrol, C. Jenkinson, D. J. M. Reynolds, D. J. Gavaghan, et al. 1996. *Controlled Clinical Trials* 17:1–12.
75. Goel, V., R. Lovlin, M. R. Lyon, R. Bauer, T. D. D. G. Lee, and T. K. Basu. 2004. *Journal of Clinical Pharmacy and Therapeutics* 29:75–83.
76. *Bundesanzeiger* (Cologne, Germany): January 5, 1989.
77. Castleman, M. 1991. *The healing herbs,* 150–154. Emmaus, PA: Rodale Press.
78. Castleman, M. 1989. *Medical Selfcare* 53:53–54.
79. Aquino, R., V. DeFeo, F. DeSimone, C. Pizza, and G. Cirino. 1991. *Journal of Natural Products* 54:453–459.
80. *Lawrence Review of Natural Products:* April 1996.
81. Wagner, H., B. Kreutzkamp, and K. Jurcic. 1985. *Planta Medica* 51:419–423.
82. Reinhard, K.-H. 1997. *Zeitschrift für Phytotherapie* 18:112–121.
83. Chang, H. M., and P. H. H. But, eds. 1987. *Pharmacology and applications of Chinese materia medica,* vol. II, 918. Singapore: World Scientific Publishers.
84. Puri, A., R. Saxena, R. P. Saxena, K. C. Saxena, V. Srivastava, and J. S. Tandonj. 1993. *Journal of Natural Products* 56 (7): 995–999.
85. Madav, S. S. K., S. K. Tandan, J. Lal, and C. Tripahti. 1996. *Fitoterapia* LXVII (5): 452–458.
86. Thamlikitkul, V., S. Theerapong, P. Booronj, W. Ekpalakorn, S. Taechaiya, T. Ora-Chom-Jan, et al. 1991. *Journal of the Medical Association of Thailand* 74:437–442.
87. Poolsup, N., C. Suthisisang, S. Prathanturarug, A. Asawamekin, and U. Chanchareon. 2004. *Journal of Clinical Pharmaceutical Therapy* 29:37–45.
88. Kulichenko, L. L., L. V. Kireyeva, E. N. Malyshkina, and G. Wikman. 2003. *Journal of Herbal Pharmacotherapy* 3 (1): 77–93.
89. Cáceres, D. D., J. L. Hancke, R. A. Burgos, E. Sandberg, and G. K. Wikman. 1999. *Phytomedicine* 6 (4): 217–223.
90. Narimanian, M., M. Badalyan, V. Panosyan, E. Gabrielyan, A. Panossian, G. Wikman, and H. Wagner. 2005. *Phytomedicine* 12:539–547.

Appendix: The herbal regulatory system

Updated and expanded by Paula N. Brown and Michael Chan
Natural Health Products Research Group,
British Columbia Institute of Technology

Drug regulation 1906–1962

Early in this century, fraud was rampant among the producers of foods and drugs in the United States. Harvey W. Wiley, then chief of the Bureau of Chemistry of the U.S. Department of Agriculture, undertook a vigorous campaign against the unscrupulous practices of both industries. His work resulted in passage of the Food and Drugs Act of 1906. Commonly but incorrectly referred to as the "Pure Food and Drug" Act, the statute was, at the time, a true innovation. This act and the 1912 Sherley amendment to it put an end to many fraudulent practices, such as the misbranding and adulteration of drugs, but failed to address effectively the problems of drug safety and efficacy.[1] Little attention was paid to such matters for more than thirty years; then, a tragedy brought drug safety concerns to the fore.

In 1937 the S. E. Massengill Company of Bristol, Tennessee, began to market an Elixir Sulfanilamide consisting of 8.8 percent of the drug in 72 percent diethylene glycol. This product, intended primarily for Southerners who liked to drink their medicines, was marketed without adequate toxicity testing in animals; in fact, it was tested only for flavor prior to marketing. Before the poisonous character of the solvent was recognized and the product could be recalled, it caused some 105 deaths from acute kidney failure.[2] The public outcry resulting from this unfortunate episode precipitated the passage of the 1938 Federal Food, Drug, and

Cosmetic Act. Among its requirements was one necessitating that drugs entering interstate commerce be proven safe. If a drug already on the market was subject to the 1906 act, it was grandfathered, and additional proof of safety was not required.

Although a number of amendments were made to the 1938 act, including a rather significant one (Durham–Humphrey) in 1951 that dealt largely with classification and procedures, it took a further tragedy to bring about substantial change. Thalidomide, a sedative drug developed by Chemie Grunenthal GmbH in Stolberg, Germany, in 1953 and widely marketed in Europe, was identified in 1962 as a potent teratogen. About 30–40 percent of the mothers who took the drug during a critical phase of their pregnancies gave birth to physically deformed babies. The drug was briefly marketed in Canada with disastrous effect. Thalidomide was never marketed in the United States, but its observed side effects caused great concern there.[3]

This period of alarm coincided with an extensive Congressional investigation of the American drug industry led by Senator Estes Kefauver of Tennessee. All of these events resulted in the passage of the Drug Amendments of 1962, often referred to as the Kefauver–Harris Amendments. These required, among other things, that all drugs marketed in the United States after 1962 be proven *both safe and effective*. Again, drugs marketed prior to 1938 were grandfathered, but every product introduced between 1938 and 1962 was subject to these requirements. This group consisted of approximately three hundred different chemical entities in about four thousand different drug formulations that were actively being sold, as well as another three thousand formulations that had been approved by the standard new drug application (NDA) process but were not actively marketed.

Implementing safety and efficacy regulations

In view of the enormity of the task of evaluating the safety and efficacy of all these products, the Federal Food and Drug Administration turned for help to the Division of Medical Sciences of the National Academy of Sciences—National Research Council. They proceeded to organize a drug efficacy study that began work in 1962. After a long series of preliminary investigations that lasted until 1969, seventeen panels, each of which concerned itself with a different therapeutic class of drugs, were formed in 1972 to examine over-the-counter (OTC) drugs. This is the category in which most herbs and herb products were found.

The panels did not utilize all sources of information because some, such as patient testimonials, anecdotal experiences of physicians, and market success of the product, were considered unreliable. Instead, they relied primarily on in vitro tests (studies not conducted in patients or whole animals) and on various kinds of clinical or patient trials. The panels themselves did not conduct these studies but, rather, evaluated information supplied to them by organizations interested in marketing the drugs.

It was obviously impossible to examine the approximately one-third of a million drug products in the OTC category, so evaluations were limited to the active ingredients in them. Because of the paucity of information resulting from in vitro tests or clinical trials with herbs, the panels were asked to evaluate a relatively small number of such drugs for specific indications. In addition, most of them had apparently been grandfathered under both the 1938 and 1962 acts and were considered immune from the effective requirements implemented at the later date.

FDA's herbal regulatory initiative

Nevertheless, the FDA undertook the regulation of herbs by what can only be termed an extremely innovative application of administrative law. The agency simply declared that any of the grandfathered drugs would be considered misbranded and subject to confiscation if any claims of efficacy were made for them that were not in accordance with the findings of one of the seventeen OTC drug evaluation panels. Thus, it became possible to sell the products only if no claims or statements regarding their value in the prevention or treatment of disease were made on the label. The word *label* was very broadly defined to include not only the words on the package but also those on any accompanying literature, such as a package insert. In this way, herbs were effectively prevented from being marketed as drugs.

Results of the FDA study of OTC drugs were released to the public in the spring of 1990.[4] Although some of the herbs, including cascara bark and senna leaf, were found to be effective as laxatives and initially placed in category I, some 258 nonprescription drug ingredients, many of them herbs, were not so judged. Of these, 142 were categorized as unsafe or ineffective (category II), and there was insufficient evidence to evaluate the effectiveness of the other 116. These were placed in category III. Again, the panels made these judgments on the basis of evidence submitted to them by the industry.

Reaction to FDA's herb regulations

Reaction of the herb industry to the FDA's stricture on therapeutic labeling of products judged ineffective by the panels took a variety of forms. Some herbs were no longer marketed. Others continued to be sold as before, apparently in the hope that an underfunded and overworked FDA might overlook, at least for a while, some of the infractions. But the most frequent course of action was simply to remove the offending information from the label and to market the product as a food, nutritional supplement, or, in some cases, a food additive. Unfortunately, this left little more on the label than the name of the herb; however, a vast literature (often laced with hyperbole and inaccuracies) was available to the consumer—ostensibly to supply product information but mostly intended to promote sales.

For some time, the FDA has maintained a list of substances "generally recognized as safe"—better known in the trade as the GRAS (generally recognized as safe) list.[5] About 250 herbs appear on this list, primarily based on their culinary use as food—that is, as flavors or spices in the culinary arts and the beverage industry. Some of them, such as ginger and licorice, are also employed for their medicinal action. Obviously, if a plant material is simply labeled "ginger," it is impossible to know whether it will be used as a flavoring or as a medicine (e.g., to prevent motion sickness). In some cases, ginger from the same package possibly would be used both ways.

The FDA formerly maintained two other lists in addition to the GRAS list. These were "herbs of undefined safety" and a list of twenty-seven "unsafe herbs." Both were flawed by inappropriate inclusions, and both were discarded in 1986 when the FDA updated its Compliance Policy Guidelines on the safety of food additives. The policy then called for herbal safety to be determined on an ad hoc basis, usually following a consumer complaint.[6] There is now no FDA compilation of unsafe or possibly unsafe herbs, but sometimes examples on these older lists are referred to in the literature.

One action that was never taken by members of the herb industry was to prove their products safe and effective and to market them as drugs. The reason for that was money; the costs involved are extremely high. A 1990 report by the Center for the Study of Drug Development at Tufts University placed the average cost for developing a new drug at $231 million and the time involved at twelve years.[7] While these figures might well be double that required for a traditional herbal remedy, the figures of $115 million and six years are still excessive.

This is particularly true considering the difficulty of obtaining a patent, as well as the exclusive marketing rights it affords, on a traditional drug that may have been used for centuries, even millennia. Unless a novel chemical entity is involved, the likelihood of patent protection is slim, and the capital investment required will likely never be recovered.

The Dietary Supplement Health and Education Act

In the mid-1980s, the consciousness level of the American public was raised considerably concerning diet, particularly the influence of food on blood cholesterol levels. In addition, controversy developed as to whether the food industry could make health claims on food labels and in advertising. This resulted in public pressure for clearer and more complete nutrition information and more rigorous evaluation of food health claims, culminating in the passage of the Nutrition Labeling and Education Act (NLEA) of 1990. This law strengthened the FDA's authority and required the FDA to adopt sweeping new regulations for food labeling.[8]

Herbal medicines and other dietary supplements that were being sold as foods were placed in a precarious position because it was difficult to apply the labeling requirements to these products. In 1991, after the FDA convened a Dietary Supplement Task Force to study this matter, the herbal medicine industry became alarmed that the task force would recommend stopping sales of amino acids, herbs, and "supplement" products. Industry-related groups reacted by mobilizing a campaign to inform Congress of their unhappiness. As a result, more protest mail was received by members of the U.S. Congress on this matter than on any other issue since the Vietnam War. Congressional response was the passage in 1994 of the Dietary Supplement Health and Education Act (DSHEA).[9,10] DSHEA defines the term "dietary supplement" as

> a product (other than tobacco) intended to supplement the diet that bears or contains one or more of the following dietary ingredients:
>
> (A) a vitamin;
> (B) a mineral;
> (C) an herb or other botanical;
> (D) an amino acid;
> (E) a dietary substance for use by man to supplement the diet by increasing the total dietary intake; or
> (F) a concentrate, metabolite, constituent, extract, or combination of any ingredient described in clause (A), (B), (C), (D), or (E).[11]

In addition to this, a dietary supplement must be "intended for ingestion in tablet, capsule, powder, softgel, gelcap or liquid form, or if not intended for ingestion in such a form not be represented for use as a conventional food or as a sole item of a meal or the diet."[11]

Despite being presented in dosage forms, dietary supplements are not considered to be within the realm of drugs. Instead, the passing of DSHEA has determined that these products are a subset of foods. With this classification, herbal products in the law's eyes are no longer medicines; rather, they are supplements, designed to promote health by supplementing the diet.

The passage of the DSHEA not only introduced the legal definition of dietary supplements. It also introduced the term "new dietary ingredient

(NDI). A new dietary ingredient is defined in DSHEA as "a dietary ingredient that was not marketed in the United States before October 15, 1994".[12]

The simple presence of a dietary ingredient in the marketplace is not sufficient for it to be considered an "old" dietary ingredient. The ingredient must have been lawfully marketed as a dietary ingredient prior to October 15, 1994 to qualify. The FDA stated that for an ingredient to be considered lawfully marketed, the manufacturer or distributor must have written evidence that the ingredient is chemically identical to a dietary ingredient legally marketed in the United States before October 15, 1994. The evidence required to prove that an ingredient was lawfully marketing must be a product invoice, bill of lading, product label or a catalogue with a date showing evidence of marketing before October 15, 1994. The FDA also made it clear that the ingredient *itself* must have been marketed, not just a formulation containing the ingredient. The mere presence of the ingredient in food is not sufficient evidence to prove prior marketing; the ingredient must be clearly marketed for its own properties.

Whether or not a dietary ingredient is considered new or grandfathered has become of immense importance these past few years because the DSHEA requires all NDIs to undergo a pre-market safety review. In the early years of the DSHEA little attention was really paid to this part of the Act. However, in recent years the FDA has stepped up enforcement of this provision. A large number of dietary supplements have become rejected or removed because they contain new dietary ingredients. In addition, many NDI submissions have been rejected by the FDA in recent years. The reasons cited by the FDA include mismatches between the marketed and tested ingredient, the use of therapeutic language, the use of food language and the use of clinical studies conducted via non-oral routes. The most often cited reasons for rejection are the lack of sufficient safety data or failure to comply with the safety standard.

The FDA will only accept data that shows the NDI in the exact form and at the dosage level described in the NDI submission. The safety evidence provided must support the NDI's use for food purposes or as an article of food. Thus, reference to traditional use as a medicine or treatment as a basis for safety is usually not sufficient. The safety evidence must clearly show the ingredient in the reference or study is identical to the ingredient in the NDI submission and at the dosage levels stated in the NDI submission.

Perhaps the most important effect of the passage of the DSHEA was permitting of claims for dietary supplement products. The DSHEA allowed dietary supplements to be advertised for their health promoting properties. As a subset of foods, they are permitted to use any claims allowed for foods in U.S. food law, except for nutrient content claims. Because they are supplements only, they are exempt from the nutrient content requirements that all other foods making health claims must follow. Dietary supplements are also allowed to make much less restrictive "structure function" claims. Dietary supplements, however, cannot claim "to diagnose, mitigate, treat, cure or prevent a specific disease or class of diseases."[13]

Structure-function claims are the most common claims made with dietary supplements. They are statements describing the product's role in affecting or maintaining structure or function in humans. The statements allowed can be made

> if the statement claims a benefit related to a classical nutrient deficiency disease and discloses the prevalence of such disease in the United States, describes the role of a nutrient or dietary ingredient intended to affect the structure or function in humans, characterizes the documented mechanism by which a nutrient or dietary ingredient acts to maintain such structure or function, or describes general well-being from consumption of a nutrient or dietary ingredient.[13]

The DSHEA also sets standards for the distribution of third-party literature. This literature, which must be of a generic nature, is available in connection with the sale of the dietary supplement, but not attached to the product; it can include articles, book chapters, or official abstracts of peer-reviewed scientific publications as long as they present a balanced view of the available scientific information. Under the law, the president of the United States appointed an independent Commission on Dietary Supplement Labels. It reviewed and evaluated label claims and statements and made recommendations to develop guidelines in the use of these claims on labels in the third-party literature.

The FDA does not require manufacturers to seek premarket approval for claims. Manufacturers need only notify the Secretary of Health and Human Services within thirty days of the first marketing of the dietary supplement and indicate the statement being used.[13] The FDA has stated that the claims, or statements, are allowed only if they are truthful and not misleading. Prohibited drug claims are also not allowed and the FDA did, and still does, review claims for that reason. Initially the FDA did attempt to review and in some cases prohibit certain claims they felt were untruthful and/or misleading, but following the decision set out in the *Pearson v. Shalala*[14] case, the FDA decided to put all responsibility for ensuring the validity of dietary supplement claims solely on the manufacturer. Although it would still prevent any drug claims, the FDA would essentially allow any structure-function claim on the label provided the label also prominently displayed, in boldface type, the following statement "This statement has not been evaluated by the Food and Drug Administration. This product is not intended to diagnose, treat, cure or prevent any disease'."[13]

The absence of pre-market approval is not limited to claims. With the exception of supplements containing NDI's the FDA does not normally require any pre-market safety or efficacy assessments on dietary supplements. Even in the case of NDI's, the premarket evaluation is only on the safety of the NDI itself not the safety or efficacy of the finished product. It was reasoned that dietary supplement use is voluntary and thus, the consumer is willing to accept some of the risk associated with such use. Thus, there is a slightly lower standard of safety for dietary supplements as compared with conventional foods. Furthermore, when the DSHEA was first passed there were no provisions that required for manufacturers or distributors to report adverse effects. Reporting of serious adverse effect for dietary supplements became mandatory for manufacturers in December of 2007.[15] Reports are submitted to FDA through the MedWatch system.

Under the DSHEA it is up to the FDA to prove a product is unsafe before it can remove it from the market or prevent its sale. The exact criteria and procedures by which the FDA proves a dietary supplement to be unsafe are major issues currently undergoing debate. In a ban on ephedrine alkaloids in dietary supplements, the FDA used a risk-benefit analysis to judge that these dietary supplements present an unreasonable risk of illness or injury under their conditions of use. The FDA's use of a risk-benefit analysis was challenged by industry members who argued that congress did not intend for such an analysis to be used for dietary supplements.[16]

The District Court for the District of Utah agreed. However, the 10th District Court of Appeals overturned the decision and ruled that Congress clearly intended to integrate a risk-benefit analysis in the qualifying of risk under the DSHEA.[16] Industry groups have challenged this ruling arguing that a risk-benefit analysis is not authorized by the DSHEA and that the Court's ruling as it stands essentially holds dietary supplements to the same standards of drugs, contrary to congress' initial intent.

In 2007 the FDA published their current good manufacturing practice (cGMP) in manufacturing, packaging, labeling or holding operations for dietary supplements in the *Federal Register*.[17] The cGMPs applies to all U.S. and foreign companies that manufacture, package, label or hold dietary supplements that are sold or offered for sale in the U.S. The cGMPs apply only to dietary supplement manufactures, and not dietary ingredient suppliers. As a result ingredient suppliers are not subject to these standards, however, it is expected that manufacturers will hold ingredient suppliers to standards that do not result in a manufacturer being in violation of CGMP.

The cGMPs are divided into sixteen subparts that describe standards for the manufacturing of dietary supplements. and was designed to ensure the quality of dietary supplements. The cGMPs explicitly state that "Quality means that the dietary supplement consistently meets the established specifications for identity, purity, strength, composition, and limits on contaminants, and has been manufactured, packaged, labeled and held under conditions to prevent adulteration..."[17]

At its core the cGMPs describe the development and implementation of a quality assurance plan. The various subparts of the cGMPs discuss topics including plant and equipment sanitation and maintenance, employee training and hygiene, process controls, dealing with GMP related complaints, labeling and packaging, record keeping and the development and use of formal standard operating procedures (SOPs).

The cGMPs place great emphasis on the establishment of specifications that ensure the identity, purity, strength and composition of finished dietary supplements products. Manufacturers must apply specifications to both finished products and incoming dietary components, including dietary ingredients. Test methods that are used to measure against specifications must be appropriate and scientifically valid. This does not mean that the tests methods have to be compendial or official test methods (although the FDA does recommend their use if they are appropriate and are available) as the paucity of official methods would make compliance impossible. The FDA will allow methods to be used provided they are suitable for their intended use and have undergone a minimum level of validation. The FDA also notes in the cGMPs that typical validation characteristics include accuracy, precision, specificity, detection limits, quantification limits, linearity, range and robustness. Various initiatives by government, industry and academia have been launched to increase the number and availability of validated methods.

The publication of the cGMPs is perhaps the most significant addition to the dietary supplement regulatory environment since the passing of the DSHEA. The cGMPs emphasize the quality assurance of dietary supplements. In recent years the quality of dietary supplements has been a significant concern among public officials and consumers. It is hoped that the implementation of the cGMPs can restore confidence in these products. However, there exist concerns that the costs of complying with these cGMPs would result in inability of many small firms and manufacturers to compete in the industry and/or significantly higher supplement costs. It is still too early to know what the exact outcomes of implementation of the cGMPs will be.

Canadian regulation of herbal products

Like the United States, Canada, has developed systems to regulate herbal products. In Canada, some herbs are considered foods, some are considered drugs and some may considered both. Where an herbal product falls is very much dependent on its history of use, its dosage and its presentation. In some cases these criteria are clearly stated, whereas in many others the dividing line is obscure.

For herbal products that are presented as therapeutic agents the classification is clear. Unlike the United States, which passed the DSHEA to regulate these products as a subset of foods capable of making mild claims, Canada took a decidedly different route and set out to regulate these products as a subset of drugs and allowed drug-like claims to be made for these products. The Natural Health Products Regulations came into effect on January 1st, 2004 and regulate all herbal remedies as well as vitamins and minerals, homeopathic medicines, traditional medicines, probiotics and other products like amino acids and essential fatty acids.[18] These products are called Natural Health Products (NHPs).

Herbs that are not presented as having therapeutic effects are not necessarily exempt from regulatory scrutiny. This is especially true for herbs that have not been commonly used as foods in Canada (which incidentally, sums up the vast majority of herbs). These herbs would be considered novel foods and would still require a very stringent safety evaluation prior to gaining market approval.

NHPs, unlike dietary supplements, are also required to undergo a pre-market safety evaluation. But, because they make claims, NHPs must also undergo quality and efficacy evaluations prior to going to market. NHPs are not, however, subject to all the provisions set out for drugs in the Canadian Food and Drug Regulations and are subject to a simplified evaluation process.[18] The authority that evaluates and regulates these products is known as the Natural Health Products Directorate (NHPD).[18]

Because NHPs are not limited to a dosage form, they can be in typical supplement forms such as capsules, pills, tablets, powders and liquids, as well as in more conventional food forms such as bars, gums, wafers or beverages. Some topical creams, lotions and toothpastes are considered NHPs. Products that require that they be delivered by puncturing the dermis (i.e. injectables) cannot be NHPs.

The definition of a Natural Health Product essentially consists of two parts: a structure component and a function component. The structure component refers to the medicinal ingredient or ingredients in the NHP, and includes "herbal remedies, traditional and homeopathic medicines and materials derived from plants, algae, bacteria fungi, or non-human animal material, amino acids, essential fatty acids, probiotics, minerals, several vitamins and synthetic duplicates of the natural ingredients".[18]

The function component is the requirement that these products are "sold or represented for the use in the diagnosis, treatment, mitigation or prevention of a disease, disorder or abnormal physical state or its symptoms in humans, restoring or correcting organic functions in humans, modifying organic functions in humans, such as modifying those functions in a manner that maintains or promotes health."[18] Simply put, NHPs can and must make drug-like claims or structure function claims. When used in NHPs, there is essentially no distinction between the two; both are considered NHP claims.

NHP claims can be divided into two categories: traditional use claims and non-traditional use claims. Traditional use claims are for products that have been used within a cultural belief system or healing paradigm for at least 50 consecutive years.[18] Unlike other jurisdictions with similar traditional use clauses, under the Canadian regulations, the cultural system is not limited to the local population. Thus, the demonstration of the traditional use of the product in another culture or country other than Canada is acceptable.

Non-traditional use claims are for those claims that do not meet the requirements of traditional use claims. These claims need to be supported by scientific evidence and the review and approval of the claim is done by the NHPD during the product licensing procedure.

Product licensing is the pre-market approval step that all NHPs must go through prior to legal sale. Following the review process, those products deemed acceptable by the NHPD in terms of safety, quality and efficacy would be given a product license and a Natural Product Number (NPN) or, in the case of homeopathic medicines, a Drug Identification Number for Homeopathic Medicine (DIN-HM).[18] The standards of evidence required to demonstrate safety, quality and efficacy will vary between products. The amount and type of evidence will be dependent on the nature of the product and the type of claim being used.

Products can consist of combinations of medicinal ingredients. The applicant is required to justify the use of such a combination and ensure that there are no safety concerns with the combination of the ingredients. Basically, four types of product license applications that can be made: a compendial application, a traditional claim application, a non-traditional claim application and a homeopathic medicine application.

Among the NHP product license application types, the compendial Application is the simplest and most restrictive. These applications are for products that contain a single medicinal ingredient that is the subject of an official NHPD monograph. For products containing more than one medicinal ingredient, the NHPD has published several product monographs that can also be used for compendial applications; however these monographs are very limited in scope and number. The NHPD has stated that as more information is obtained and reviewed, more monographs will become available for use.

The monographs provide specific details including, but not limited to, the medicinal ingredient's proper name, common name, and source. Additionally, the monographs specify the route of administration, the dose, quality specifications for the product, and duration of use for any appropriate subpopulation. Products that are applying for a compendial application must conform to these specifications.

The monographs also indicate a recommended use or purpose for the ingredient, as well as risk information associated with the ingredient. An applicant may make alterations to the wording of these elements, but cannot alter their meaning. If the product matches the required criteria as set out in the monograph, a compendial application can be made and the monograph can be used as the sole source of evidence for the safety and efficacy of the product. NHPD has stated that compendial products will be reviewed within 60 days of the date upon which the directorate receives a complete submission.[18]

As noted above, traditional claims need to refer to the medicinal use of the product for 50 consecutive years. To qualify for a traditional use claim, the product needs to be prepared in the traditional manner described in the submitted evidence.[18] The evidence must include at least two independent references (i.e. references that do not cite the same source, or each other, as the main source of information regarding the traditional use) that support the recommended conditions of use or one acceptable pharmacopoeia reference.

Applicants for a nontraditional claim must submit a nontraditional claim application. The amount of evidence needed to for these applications will vary with the product. At a minimum, two independent pieces of evidence supporting the product's safety and efficacy is required. The NHPD places importance on the type and level of evidence submitted in the application. To obtain a license, the evidence submitted must support the safety of the product and the claim that is being made. Different types of claims would require different levels of evidence.

All manufacturers, packagers, labelers and importers of natural health products are required to hold a site license. The site license will specify exactly which of the previously listed activities the applicant is allowed to do with respect to NHPs. In order to obtain a site license the applicant must demonstrate that they are in compliance with the good manufacturing practices (GMP) requirements set out in the regulations.

The NHP regulations are still relatively new. Health Canada and the NHPD are currently doing the first major review of the system. They are looking over the results and issues that have surfaced since the implementation of the regulations.

Among the issues to address is the large delay experienced in the processing and issuing of product licenses. A major component of the delay is the time required to process the backlog of license applications created by implementation of a new regulatory regime. The NHPD started on day one with zero licensed products and as many as 40,000 applications. When the backlog is completed, action on applications is expected to proceed in a more timely fashion. The NHPD has implemented and continues to implement procedures it believes will help streamline the process. Some of these proposed procedures aim to address some of the concerns that manufacturers have brought forth. These include the need for better guidance and information on exactly what type and amount of evidence applications required to substantiate claims and increasing the quality and quantity of published monographs. The NHPD has also introduced new on-line ingredient and product databases. These databases provide limited information on both ingredients and products that have been approved for sale. As more information is gathered the NHPD hopes to significantly expand the information available in the databases. A major innovation that is hoped to significantly improve the efficiency of product license reviews is the use of the NHP On-line System. This system, which is currently undergoing pilot studies, will allow for the support of electronic preparation and submission of license applications. It is hoped that through the implementation of these initiatives that product licensing process can be improved in terms with respect to time, efficiency and costs.

Another major concern among manufacturers and consumers is the proposed fees associated with the regulation of NHPs. The NHPD had stated that user fees would be required to cover a portion of the costs of its operation. So far, no fees have been charged. Recently Health Canada has set out a proposed cost recovery program that has drawn concern and anger among manufacturers and consumers. The proposed program is being heavily debated and the eventual outcome is anxiously awaited.

The uncertain future of herbal product regulation

At the time this is written, many questions on the regulation of herbal products in the United States and Canada still remain. Both jurisdictions are going through an important transition phase that will greatly influence the future of herbal product.

The years since the DSHEA was first passed has seen an environment where manufacturers have had a relatively free hand with little interference from regulators. The last few years have seen a shift away from this laissez faire mentality. Complaints and concerns from consumers concerning the safety and efficacy of dietary supplement products have resulted in more regulatory scrutiny. The FDA has increasingly used the NDI provision to force pre-market approval. It has also begun to develop a risk analysis system that can use to prohibit the sale of products the organization perceives to be threats to health. In addition, the provisions outlined in the new cGMPs for dietary supplements have significant effects on the industry.

In Canada, the first major review of the still relatively new NHP regulations is being carried out. The transition into this regulatory scheme has not been a smooth one and many issues still need to be addressed. Foremost among these is the time required for product license reviews and the ramifications the proposed cost-recovery program will entail.

It is clear that consumers want access to these products. But the age-old questions on how a regulatory system can ensure quality, efficacy, safety and accessibility to these products still remain unanswered. Although it is unclear whether the current initiatives in these two countries will ultimately answer these questions, one can feel sure that these initiatives will have a considerable effect on the herbal product industry for many years to come.

References

1. Ziporyn T. 1985. The Food and Drug Administration: how those regulations came to be. *Journal of the American Medical Association* 254(15):2037-2046.
2. Leech P.N. 1937. Sulfanilamide – a warning. *Journal of the American Medical Association* 109:1128.
3. Botting J. 2002. The History of Thalidomide. *Drug News Perspective* 15(9):604-611.
4. Blumenthal M. 1990. FDA declares 258 OTC ingredients ineffective: many herbs included; prunes are not an effective laxative, says FDA panel! *Herbalgram* 23:32-33.
5. Winter R. 1984. *A consumer's dictionary of food additives,*rev. ed., 8-9. New York: Crown Publishers Inc.
6. Israelsen L. 1986. Major FDA policy shift on herbs. *HerbalGram* 10:1-2.
7. DiMasi J.A. 1992. Cost of innovation in the pharmaceutical industry. *Pharmacoeconomics* 1(S1):13-20.
8. Barrett S. 1993. New food labeling regulations issued. *Nutrition Forum* 10(4):25-30.
9. Blumenthal M. 1994. Congress passes dietary supplement health and education act of 1994: herbs to be protected as supplements *HerbalgGram* 32:18-20.
10. Israelsen L. 1995. Phytomedicines: the greening of modern medicine *The Journal of Alternative and Complementary Medicne* 1(3):245-248.
11. 21 U.S.C. Sec. 201 (2004).
12. 21 U.S.C. Sec 413 (2004).
13. 21 U.S.C. Sec. 403 (2004).
14. *Pearson v. Shalala,* 164 F. 3d 650 (D.C. Cir. 1999).
15. Dietary Supplement and Nonprescription Drug Consumer Protection Act of 2006, Pub. L. no. 109-462, 120 Stat 3439 (2006).
16. *Nutraceutical v. Von Eschenbach,* 459 F. 3d 1033 (10th Cir. 2006).
17. Current Good Manufacturing Practice in Manufacturing, Packaging, Labeling, or Holding Operations; Interim Final Rule, 72 *Federal Register* 34751-34958 (2007).
18. Natural Health Products Regulations (2003), SOR/2003-196.

Index

N

P